The Case for Grassroots Collaboration

The Case for Grassroots Collaboration

Social Capital and Ecosystem Restoration at the Local Level

John Charles Morris, William Allen Gibson, William Marshall Leavitt, and Shana Campbell Jones

LEXINGTON BOOKS
Lanham • Boulder • New York • Toronto • Plymouth, UK

Published by Lexington Books
A wholly owned subsidiary of The Rowman & Littlefield Publishing Group, Inc.
4501 Forbes Boulevard, Suite 200, Lanham, Maryland 20706
www.rowman.com

10 Thornbury Road, Plymouth PL6 7PP, United Kingdom

British Library Cataloguing in Publication Information Available

Library of Congress Cataloging-in-Publication Data

Morris, John C. (John Charles), 1959-
The case for grassroots collaboration : social capital and ecosystem restoration at the local level / John Charles Morris, William Allen Gibson, William Marshall Leavitt, and Shana Campbell Jones.
p. cm.
Includes bibliographical references and index.
ISBN 978-0-7391-7696-2 (cloth : alk. paper) -- ISBN 978-0-7391-7697-9 (electronic)
1. Restoration ecology--Chesapeake Bay Region (Md. and Va.) 2. Restoration ecology--Citizen participation. I. Title.
QH104.5.C45M67 2013
639.909752--dc23
2013018790

™ The paper used in this publication meets the minimum requirements of American National Standard for Information Sciences Permanence of Paper for Printed Library Materials, ANSI/NISO Z39.48-1992.

Printed in the United States of America

This book is dedicated to the tens of thousands of citizens who give freely of their time and money to make their communities cleaner, better places.

Contents

List of Figures ix
List of Abbreviations xi
Acknowledgments xiii
Foreword xv
Introduction xix

1 Grassroots Collaborations to Solve Environmental Problems 1
2 A Framework for Collaboration: Building Social Capital, Improving Environmental Outcomes, and Maximizing Democratic Values 19
3 The Chesapeake Bay: Local Efforts, Regional Connections 53
4 Restoring Princess Elizabeth's River 85
5 The Return of the Oyster to the Lynnhaven River 143
6 The Nansemond River Preservation Alliance: A Tale of Two Rivers 183
7 Conclusion: One Watershed at a Time 217

Appendix 233
References 237
Index 251
About the Authors 257

List of Figures

Fig. 1.1	The Chesapeake Bay Watershed	2
Fig. 2.1	A Model of Collaboration	21
Fig. 2.2	An Enhanced Model of Collaboration	47
Fig. 3.1	The Hampton Roads Region	56
Fig. 4.1	The Elizabeth River Watershed	88
Fig. 4.2	Organization Chart—Elizabeth River Project	112
Fig. 5.1	The Lynnhaven River Watershed	144
Fig. 5.2	Organization Chart—Lynnhaven River NOW	162
Fig. 6.1	The Nansemond River Watershed	185
Fig. 6.2	Organization Chart—Nansemond River Preservation Alliance	200

List of Abbreviations

BMP	Best Management Practices
CBF	Chesapeake Bay Foundation
CBP	Chesapeake Bay Program
CBPA	Chesapeake Bay Preservation Act
CCWC	Choose Clean Water Coalition
DCR	Virginia Department of Conservation and Recreation
DEQ	Virginia Department of Environmental Quality
EPA	US Environmental Protection Agency
ERP	Elizabeth River Project
HRPDC	Hampton Roads Planning District Commission
HRSD	Hampton Roads Sanitation District
LRN	Lynnhaven River NOW
LRRT	Living River Restoration Trust
NFWS	US National Fish and Wildlife Service
NOAA	National Oceanographic and Atmospheric Administration
NRPA	Nansemond River Preservation Alliance
ODU	Old Dominion University
TMDL	Total Maximum Daily Load
USACE	US Army Corps of Engineers (also abbreviated as "the Corps")
VCN	Virginia Conservation Network
VEE	Virginia Environmental Endowment
VIMS	Virginia Institute of Marine Science
VPA	Virginia Port Authority
WIP	Watershed Implementation Plan

Acknowledgments

As with any project of this size, the final product is the result of the work (and cooperation) of a great many people. First, we wish to acknowledge the tremendous cooperation and enthusiasm of the executive directors of our case organizations: Karen Forget of Lynnhaven River NOW, Marjorie Mayfield Jackson of the Elizabeth River Project, and Elizabeth Taraski of the Nansemond River Preservation Alliance. These three people were remarkably generous with their time, expertise, and knowledge, and were instrumental in identifying and contacting our initial set of interviewees. We also wish to thank a group of PhD students at Old Dominion University who provided technical and logistical support for the project. Stephanie Joannou helped with interview transcriptions, editing, and final production. Luisa Diaz-Kope participated in interviews, provided editing support, and was instrumental in preparing the literature review for this project. Matthew Gable also participated in interviews, created the figures for the final manuscript, and provided invaluable technical assistance throughout. Katharine Neill provided critical editing and support work for us. Adale Martin also provided additional assistance in editing and production. Though their names do not appear on this volume, their contributions were critical to its timely completion.

Thanks are also due Professor Gilbert Yochum, dean of the College of Business and Public Administration at Old Dominion University, for funding in support of our interview transcription costs. We also thank Dr. Tancy Vandecar-Burdin, associate director of the Social Sciences Research Center at Old Dominion University, and her staff—Tarah Gibbs, Sarah Resnick-Leutke, and Wendi Wilson-John—for their outstanding transcription services. Their professionalism, responsiveness, and ability to maintain good humor while keeping track of dozens of audio and text files are very much appreciated. We are also indebted to our editor, Justin Race, and his assistant, Alissa Parra, without whom this manuscript would be a much less perfect product.

We also wish to extend special thanks to the dozens of citizens, staff members, agency personnel, and others who gave so freely of their time to talk with us and share their insights, expertise, stories, and observations. These men and women were unfailingly responsive to our requests for interviews, and could not have been more engaged and helpful. This book would not have been possible without their interest and cooperation, and we hope this work conveys the remarkable efforts and successes these people have enjoyed.

Finally, we wish to thank our families for their love and support during this project.

Foreword

Our nation has made great strides in protecting and restoring our environment, but the landscape of environmental policy in the United States is in transition. Several decades of command-and-control, regulatory frameworks created by the federal government in the 1970s have done much to protect our natural resources from industrial and municipal pollution, but they are reaching the limits of their intended capacity. We are now at a point where considerable individual choices must be made with regard to our environmental impacts, and decisions over how we develop our lands at the local government level are paramount. These actions simply do not fit well within the historical statutory and regulatory framework. Meanwhile, the federal government's role in protecting the environment is currently the focus of great discussion. Philosophies and political positions about the role of government are being challenged and economic resources, while historically significant, are trending down. Some continue to believe that we must choose between environmental protection and economic growth, others argue that successful command-and-control approaches simply need to be updated to meet today's challenges, while still others urge that innovative market-based approaches are more efficient than traditional controls.

Regardless of your position—we are at a crossroads. New energy, new partnerships, and new ideas are needed.

The cause and severity of environmental problems are oftentimes easy to quantify, but exceptionally difficult to solve. While our scientific knowledge of sources of pollution and their effects on health and the environment is immense, our ability to require—or even encourage—desirable behaviors from businesses and individual citizens alike is more limited. To complicate the matter, pollution does not respect political boundaries drawn on a map. Excess nitrogen entering the Susquehanna River in New York State does not stay in New York; it is carried downstream through Pennsylvania and Maryland to the Chesapeake Bay, where it washes further down the Bay across the Virginia line, and ultimately to the Atlantic Ocean. Pollution is not a local problem or a state problem; at the very least, it is a national problem. Certain kinds of pollution can have adverse impacts thousands of miles away, and in doing so become an international problem. Yet controlling pollution starts where it begins: at the local level.

At the same time, Americans are more aware of environmental challenges and the effects of an unhealthy environment than at any time in our history. Countless studies indicate broad support for a clean environment and many Americans are concerned with long-term damage to the Earth's ecosystem. These concerns cross ideologies and political parties, as well as income and education categories, state of residence, and many other factors. We are thus caught in something of a quandary: Americans want a clean, healthy environment, but they are also concerned about economic impacts and the appropriate role of government. As the challenges mount and resources are further constrained, the need for alternative approaches grows concomitantly.

The Chesapeake Bay is one of the great natural treasures of the United States. In the dawn of the seventeenth century, Captain John Smith, founder of the Jamestown Colony, wrote eloquently of the natural splendor and plentiful bounty of the Bay. The largest estuary in the United States, the Chesapeake Bay has been known for its production of oysters, blue crab, Rockfish (Striped Bass), and other species. Unfortunately, a combination of overfishing, pollution and disease has all but ended commercial oyster fishing in the Bay, the blue crab population teetered on the precipice of collapse for many years, and a complete moratorium on the harvest of Rockfish in the mid- to late 1980s was needed to rescue the iconic fish.

The Bay is also under immense and constant pressure from an increasing human population. The Chesapeake Bay watershed is home to more than 17 million people living in six states and the District of Columbia. Without additional efforts to reduce pollution, as population continues to expand, so too may the impacts from agricultural and residential fertilizers, industrial pollution, wastewater-treatment plants, and land development.

This book takes a fresh look at efforts to address pollution in the watershed. After decades of effort by talented and committed people in national, state, and local governments, a great deal of progress has been made, but we remain far short of our Chesapeake Bay–restoration goals. Regional groups especially have helped greatly to raise awareness of the environmental challenges in the watershed, but even their efforts cannot engage everyone.

A more recent approach involves the mobilization of citizens in communities to address water-quality issues in their local area, bolstering regional and national efforts. Groups of concerned citizens are working with others in their communities to raise awareness, educate, and transition personal behaviors to more environmentally friendly approaches. These grassroots groups are not concerned with the entire Bay; they are concerned with the waters immediately around them. Water-quality successes in each of these small tributaries contribute positively to water quality in the Bay. The larger "Bay" will only be restored if local efforts are successful—stream by stream, river by river.

Full restoration of the Chesapeake Bay is not the responsibility of these citizen groups. As the authors of this book point out, however, there is a critical role for these groups, and the incorporation of grassroots environmental organizations into a broader strategy of Bay restoration can mobilize tremendous resources to address the larger problems. These local organizations have shown that collaborating with governments, businesses, and other stakeholders can bring new, innovative resources and approaches to bear on environmental problems, and improve upon traditional regulatory approaches.

By combining national and local efforts, the Chesapeake Bay can once again become a healthy ecosystem. That end goal is within our reach and I sincerely thank the many dedicated grassroots organizations who are working tirelessly to ensure our success.

Jeff Corbin
Senior Advisor for the Chesapeake Bay and Anacostia River
US Environmental Protection Agency
Washington, DC

Introduction

The question of how best to address seemingly intractable societal problems is one that has received a significant amount of attention from scholars and policy makers alike. For the past twenty years, much of this attention has been focused on the use of collaboration as the "Holy Grail" of policy tools to achieve desirable social goals. Environmental problems often seem particularly well-suited to collaborative approaches, in that they are often multijurisdictional, expensive, and require the extensive application of scientific expertise that often outstrips the ability of policy makers to comprehend the nuances of the information produced, as well as the alternatives available to address the problem.

Our anecdotal observation is that both scholars and policy makers tend to focus on collaborative arrangements consisting of existing institutional-level actors—agencies and bureaus, local governments, and large advocacy groups. Oftentimes, these collaborative efforts are applied at a macro level, such as the entire Chesapeake Bay estuary. As Howard Ernst (2010) has pointed out, collaboration at this level has been something less than effective as a means to address water quality in the Bay. On the other hand, very little attention has been directed toward more locally focused efforts. The genesis of this project was a chance conversation at a conference in 2009 at which three of us presented papers on Lynnhaven River NOW and the Elizabeth River Project. These conversations led to more serious discussions about the remarkable work these organizations were doing, and the underlying elements that contributed to their success. This book is the direct result of those discussions.

Those initial discussions were focused around a curious observation: our cases represented citizen-based efforts to address significant water-quality issues in their respective watersheds. Both organizations had been formed by concerned citizens, and both were actively partnering with important stakeholder groups in their watershed. More importantly, both organizations espoused the use of collaboration, and overtly rejected confrontational, litigious techniques. Environmental protection is not a policy arena one often equates with collaboration, nor is it an arena known for success when driven by grassroots citizen efforts. Most importantly, both organizations had enjoyed success in addressing water-quality issues in their respective watersheds. Finally, it was notable that there were two such organizations located in different watersheds, but physically within a few miles of each other.[1]

At the same time, we noted some important differences between the two organizations. First, one organization was focused on a watershed that was a historically important industrial river, with literally centuries of environmental abuse to address. The other river had historically been an important oyster-production area, but had fallen victim to poor water quality due to the nonpoint runoff from a burgeoning residential population. One river separated or ran through four different local government jurisdictions; the other river was wholly located within a single political jurisdiction. The major stakeholders for one river included shipyards, cargo terminal operators, chemical companies, and the US Navy; the stakeholders on the other river were largely residential property owners.

With both similarities and differences in mind, we began to think about ways to better understand how these two different groups accomplished their goals. The answer became clear almost immediately: collaboration. An early exploration of the literature revealed that while much had been written about collaboration, there was little research that focused specifically on citizen-based, grassroots collaboration. As our ideas began to take shape, we also realized that the key to understanding citizen-based collaboration was social capital. Much like the concept of legal-rational authority in Max Weber's bureaucratic model, social capital is the "coin of the realm" for successful collaboration.

This revelation initially seemed to raise more questions than it answered. For example, the political culture of the region is an amalgamation of individualistic and traditionalistic cultures (see Elazar 1984), not a combination that seemed to lend itself to altruistic collective action. Yet these organizations not only existed, they were thriving. More local investigation revealed several other similar organizations in the area doing much the same things as our two initial cases. In fact, we initially identified a total of ten citizen-based groups just in the Hampton Roads area, all in different watersheds (or tributaries to watersheds), who espoused collaborative approaches to address their goals.

The Hampton Roads region of Virginia, historically known as Tidewater Virginia, is located in the extreme southeastern corner of the state. To the north lies the Chesapeake Bay; the Atlantic Ocean lies to the east; North Carolina lies to the south; and rural farmland lies to the west. The region is home to the Jamestown Colony, founded in 1607, as well as Yorktown, where the British ended their quest to quell the American Revolution in 1781. The region is also home to the oldest naval shipyard in the country, and maintains a significant military presence even today—it is home to the world's largest active Navy base, and all five uniformed services (Navy, Army, Air Force, Marines, and Coast Guard) have substantial facilities in the region. Norfolk is also a deep-water port, and is thus the site of a thriving cargo port. Water dominates the region in many respects—it not only defines the geography of the area, it also defines the economic, cultural, and historic foundations of the region.

MAJOR THEMES OF THE BOOK

There are five major themes explored in this volume. These themes are derived from the analytical framework presented in chapter 2, and are supported through our three case studies as well as our exploration of the broader Bay-wide context. The first of these themes is that voluntary environmental collaboration can make a difference in the environmental quality of the watershed. These groups are able to muster stakeholders and resources in ways that enhance environmental outcomes. Although the actions of the groups themselves do not always have a direct, measurable effect on the environment, they can act as catalysts to encourage or facilitate the actions of their collaborative partners. In other words, these efforts can result in actions that would not have happened but for the actions of these groups.

A second theme concerns the centrality of social capital in grassroots collaborative efforts. Successful collaboration requires trust, and trust is both derived from, and a necessary precondition of, social capital. Moreover, to the extent the actions of these groups result in tangible results, these groups can enhance the amount of social capital in a community and show citizens that their time and effort makes a difference. The ability of these groups to both mobilize existing social capital and build additional social capital in a community is key to successful collaboration.

A third theme is a commitment to place. Citizens are motivated to make a difference where they live, and personal history is also an important factor. It may be hard to motivate someone to care about a problem hundreds of miles from her home, but it is relatively easy to convince someone to care about the river she sees every day in her back yard, or about the river for which she has fond childhood memories. Moreover, these people are likely to share that motivation with other people physically close to them—their neighbors may live on the same river, and people may be more likely to trust their neighbors to act in good faith.

The fourth theme we explore is that of context—in short, context matters. The issues facing the rivers in each of our three case studies are different, and the ways in which the different groups go about their work is also different. Moreover, the nature of the community setting—rural, suburban, urban, single- or multijurisdictional—can also determine the nature of the collaboration. An important component of our analysis relates to Daniel Elazar's (1984) concept of political culture, and the ways in which the political culture of an area helps determine the kinds of action possible within the setting. The ways in which citizens relate to their government help determine the kinds of actions possible, and help to define the structure and operation of the collaborative effort. As we will see, this issue takes on greater importance with the implementation of

Total Maximum Daily Load (TMDL) requirements in watersheds across the United States.

Finally, a fifth theme of this work explores collaboration between governments and nonprofit organizations. Collaboration is a tool that allows communities to address "wicked problems" (Rittel and Weber 1973) that would otherwise go unaddressed. Lester Salamon (2002, vii) notes that scholars have failed to examine "the extent to which public problem solving has come to embrace the collaborative actions of governments at multiple levels and both government and private institutions." Salamon's concern is that public managers are not equipped to address cross-sectoral efforts; our cases highlight some of the challenges and opportunities for both public managers and citizen collaborative groups. Finally, partnerships between governments and grassroots citizen organizations can serve to provide critical legitimacy to public goals. In an era in which governments are under fire from many quarters, these partnerships can enhance the legitimacy of governments. Likewise, government's willingness to partner with citizen groups can enhance the legitimacy of these citizen-based efforts, and help convince other potential partners (e.g., private-sector actors) that such partnerships are worthy of their attention and participation.

A recent article in *Public Administration Review* suggests that collaboration is driven by one of two conditions. The first condition is that organizations decide that collaboration is the best possible answer to achieve their goals. The other possibility is that organizations fail into collaboration—they are unable to accomplish a task on their own, and this failure leads to efforts to seek additional resources from other organizations (Bryson, Crosby, and Stone 2006). We believe there is a third condition, one overlooked in the literature. We suggest that organizations can choose collaboration willingly, because collaboration is consistent with a broader set of communitarian values. These values were highlighted by Alexis de Tocqueville (1835 [1997]) nearly two hundred years ago, and still serve to define American culture in the new millennium.

ORGANIZATION OF THE BOOK

The first three chapters of this book establish the context and basis for the analysis of three distinct grassroots environmental collaborations working in close geographic proximity. The following three chapters present the analyses of each of the case studies, and the final chapter presents the conclusions and the lessons learned from the cases.

Chapter 1 examines the process and outcomes of environmental collaborations in order to draw generalizations about how they are organized and managed so as to appeal to nonstate actors to collaborate effectively where competing interests and the economic primacy of the public-policy process generally create a competitive and hostile environment for environmental politics. This chapter examines the types and magnitude of environmental problems that grassroots collaborations seek to resolve, how grassroots collaborations are structured, and how linkages are established and maintained between public, private, and nonprofit participants. We also present our research questions to be addressed in the study.

Chapter 2 presents an academic framework through which to study grassroots collaborations, and will be most relevant to those with an interest in our theoretical perspective. We employ Easton's (1965) systems model as a heuristic to frame the elements of analytical interest. The chapter introduces concepts and themes important for understanding how and why environmental collaborations operate the way they do and that are also important for the measurement of success, and explains how these elements are linked together. The grassroots collaborative framework includes the following:

- Political culture. The Hampton Roads region appears to be a mix of traditionalistic and individualistic political cultures described by Elazar (1984), which would tend to diminish the chance of success for grassroots collaborations. McNamara's (2010) work on Virginia's Coastal Zone Management Program argued that it is possible to build on and take advantage of some features of traditionalistic cultures in employing collaborative techniques.
- Material resources. The ability to attract and distribute tangible resources is critical for undertaking environmental remediation efforts.
- Organization structure. The way collaborations are structured has been a major variable for the process and success of multisector collaborations (Thomson and Perry 2006).
- Environmental outcomes. The most important question to ask about grassroots collaborations is not the quality of the collaboration process and the ability to make decisions, but the extent to which they lead to improved environmental outcomes, measured by direct, observable changes in environmental conditions.
- Social capital. Developing levels of trust and willingness to civically engage is critical work for any collaboration process, particularly in a political environment that has been characterized in the literature as traditionalistic.

The last section of chapter 2 also includes a brief discussion of the methods employed in the study.

Chapter 3 tells the story of the unique political economy and various uses of the three rivers and describes their respective levels of environmental deterioration. This chapter provides the wider political and environmental context in which the three grassroots organizations analyzed in this book operate. The chapter also examines efforts to link our three case-study organizations into region-wide collaborative interactions to address cross-boundary problems and enhance social capital. The linkage of these smaller, grassroots collaborative organizations to each other and to umbrella environmental organizations provides them with a stronger voice to impact policies and practices that affect the individual river watersheds.

Chapters 4, 5, and 6 analyze three cases of grassroots collaborative organizations that work and exist in close geographic proximity to implement ecosystem restorations or environmental preservation on three watersheds that drain into the southern Chesapeake Bay. The question of how organizational and environmental challenges are managed is central. Some of the particulars addressed include the conditions leading to collaboration, how each organization is structured and has evolved, how the participation of competing stakeholders with a history of difficult relations or no relations at all is achieved and sustained, barriers to success and how those barriers are managed and overcome, and innovative actions undertaken to achieve successful implementation of ecosystem restorations. We pay considerable attention to factors that differentiate successful from less successful management of the collaborations, on the one hand, and factors that differentiate successful from unsuccessful environmental outcomes, on the other, within an environment of complex intergovernmental relations, limited fiscal resources, and competing claims on each of the participating stakeholders.

Chapter 7 summarizes the major findings of the research, and revisits the themes discussed in this introduction. We also identify factors that can provide guidance for organizing and sustaining grassroots organizations to achieve successful program and policy outcomes throughout the Chesapeake Bay estuary and other regions of the nation.

This book offers an in-depth examination of the structure, processes, goals, and effectiveness of grassroots environmental organizations. Debates on the national political scene drift from philosophical positions on the role and scope of government, "dark green" versus "light green" approaches to environmental policy (Ernst 2010), and questions of who should bear the cost of both environmental protection and pollution remediation. This book addresses three groups that have set aside the estuary-wide policy questions in favor of direct action in their communities. The ability of these collaboratives to achieve meaningful outputs and outcomes may serve to offer new solutions to environmental remediation, and generate a new thread in the national environmental-policy dialogue.

NOTE

1. Are the time of these initial conversations, our third case study, the Nansemond River Preservation Alliance had not yet been formed.

ONE

Grassroots Collaborations to Solve Environmental Problems

It is better to work with *than* against *one another.*

—Eric Poncelet (2004)

Efforts to address water quality in the Chesapeake Bay have been frustratingly ineffective. Over the years, a series of initiatives, laws, agreements, and compacts have been enacted and implemented, but the Bay and many of its tributaries remain significantly impaired. Billions of dollars have been appropriated and spent to address water quality, and millions more have been spent gathering scientific information on the watershed. In spite of the best-intentioned efforts of generations of policy makers and scientists, little progress has been made toward measureable improvement in the water.

Many of these efforts have centered on either federally led efforts to address pollution in the Bay, or state-level (and interstate) efforts to develop and enact programs. The success of interstate efforts is particularly disheartening; while the Bay itself only borders two states (Maryland and Virginia), four other states and the District of Columbia are included in the watershed via rivers that empty into the Bay (see figure 1.1).

This book examines a different approach to Bay restoration. Rather than relying on large governmental structures to address water quality, or large-scale cooperative structures made up of different governments, nonprofit organizations, and interest groups, all trying to address the Bay as a whole, the organizations presented in this book focus on small, limited tributaries to the Bay. These groups are not large in membership; they function on small budgets, and they are not interested in confrontational, litigious strategies. They do not employ large staffs of experts, and they do not occupy lavish offices. Instead, they are groups founded and sup-

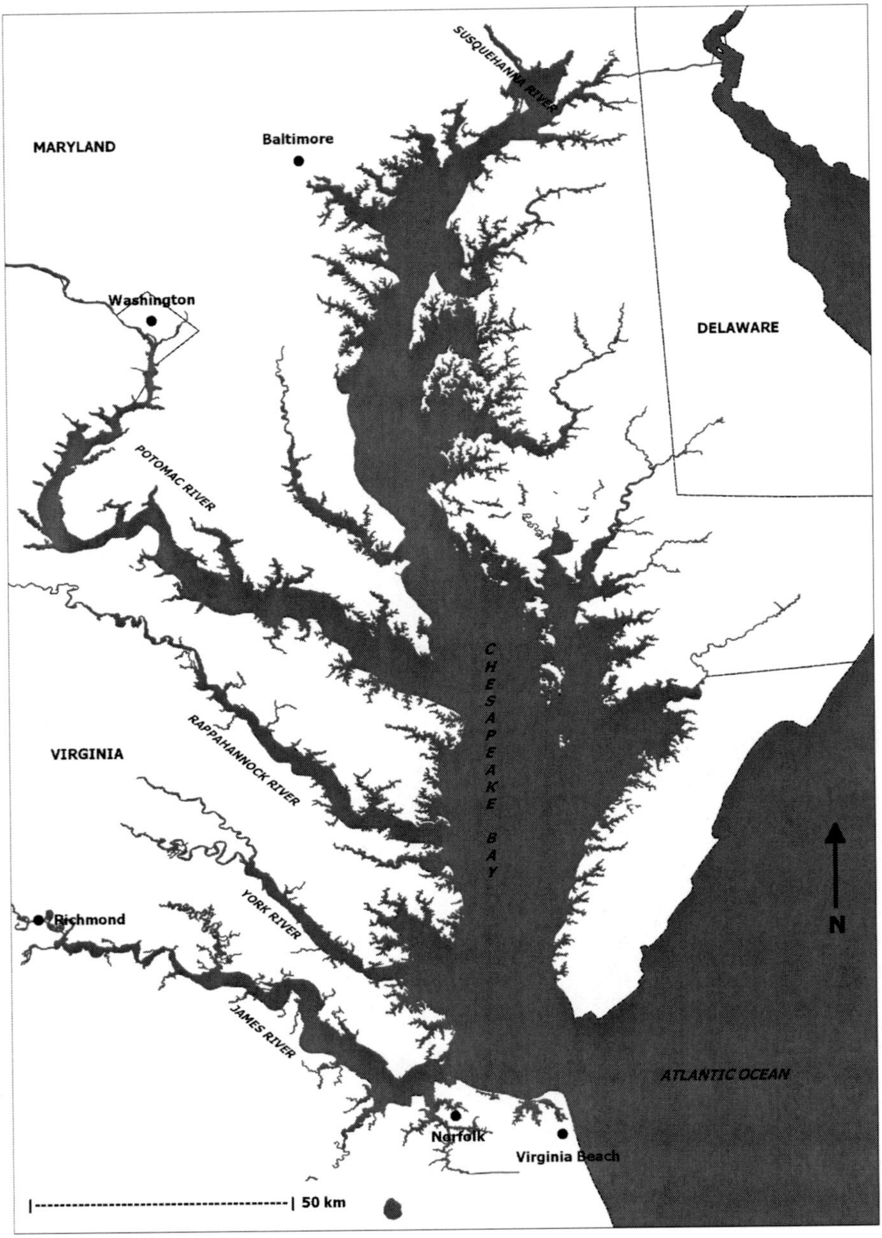

Figure 1.1. The Chesapeake Bay Watershed

ported by concerned citizens in the community who employ cooperative and collaborative techniques to address water quality issues in their communities. They are interested in "what works," and they employ a combination of trust and communitarian spirit in the tradition noted by Alexis de Tocqueville (1835 [1997]) nearly two hundred years ago.

Our research presents the stories of three such organizations, all located in the extreme southeastern corner of Virginia, in a region known locally as Hampton Roads. Our purpose is to better understand how these groups form, how they operate, and how they draw on the resources of their communities to address seemingly intractable water-quality problems. These groups are notable not for their mere existence, but for their successes—each of these groups has had a measurable impact on its watershed, and has increased social capital in its community in the process. The groups' use of collaboration to accomplish their goals, built on a solid foundation of social capital, is the key. Although there have been Bay-wide collaborative efforts to address water quality, the difference in scale between Bay-wide efforts and local watersheds is an important variable.

THE GROWTH OF COLLABORATIVE APPROACHES TO ECOSYSTEM RESTORATION

Top-down, command-and-control, regulatory approaches favored by governments and environmentalists beginning in the late 1960s brought significant improvements in air and water quality following the landmark legislation of the late 1960s and early 1970s. Yet the environment continues to be degraded, existing problems get worse, and new problems emerge. Today the problems are generally too complex, expensive, and politically divisive for one agency to solve or manage on its own. The national political environment itself has a hard-edged partisan, ideological flavor that makes it unlikely that Congress will fashion new policies and implement programs to solve the most intractable problems, or even those problems that are regional and clearly defined. The differences between the two national political parties about the role, size, and cost of government have deepened in the past twenty years and severely narrowed the range of available policy and program options, effectively ending the exclusive federal role as "chief problem solver." Even available environmental regulatory options based in existing legislation are frequently unproductive because they engender protracted negotiation and numerous legal challenges or are circumvented in the administrative regulatory process. In the past, significant environmental events, such as oil spills, drove the national agenda and led to major environmental legislation. However, it appears unlikely that such catastrophic events would result in a similar focus in today's political climate. Environmental problems in the early decades of the twenty-first century will rarely be addressed or solved by a singular, federally mandated regulatory approach, managed by one federal agency.

As recently as 1990, nongovernment actors and academicians still, by and large, looked to the federal government to develop policy solutions and implement them through the regulatory process. That approach began to shift in the early 1990s. John DeWitt (1994) described three early statewide efforts, launched in the early 1990s, to address difficult local problems, albeit problems with national implications (agricultural chemical runoff, restoring the Everglades ecosystem, and electricity conservation), through nonregulatory, participatory, and consensus-based approaches. New ideas and tools began to shape the response of government and nongovernment stakeholders as governments made efforts to enlist private sector and not-for-profit, nongovernment actors in multisector, collaborative approaches to define problems, propose solutions, and implement programs to achieve improvements to the environment. The use of citizen action had become a new tool favored by governments at all levels for enhancing environmental policy implementation as the 1990s drew to a close. The recent emergence of a new type of collaborative effort, organized, led, and managed by voluntary grassroots environmental collaborations, in particular poses unique challenges and opportunities for environmental politics.

Why This Research Matters

Government agencies have been the primary managers of consensus-based environmental collaborations. Substantial research exists about why environmental collaboration occurs and about the process itself, though that research generally addresses government-organized collaborations (see, for example, Sabatier, Weible, and Ficker 2005; Leach 2006; McGuire 2006). A different type of collaborative approach has been developed and used frequently in recent years. These collaborations are organized, led, and managed by grassroots organizations established specifically to address a particular set of environmental problems within limited geographic boundaries.

As yet, no clear, research-based understanding has been developed as to how the motivation, processes, strategies, and institutional development differ from government-led collaborations. Social capital (Lake and Huckfeldt 1998; Putnam 2000) appears to be an important element for collective action and institutional performance in collaborative efforts (Adger 2003). The process of developing the social capital, the trust and willingness to civically engage, is important because, whereas government-organized collaborations begin the process with some level of existing social capital, or at least institutional standing in the policy process, grassroots-organized collaborations generally start with little social capital and with no institutional standing. Heightened levels of politically relevant social capital tend to enhance the likelihood of continued participation by members of a collaborative process and increase the opportunities for success. Grassroots collaborations must develop that social capital and establish themselves as credible participants in the policy process. It is important to learn more about how grassroots organizations, beginning with none of the institutional standing of government-organized collaborations, develop and maintain the social capital that is most likely a necessary condition for success in environmental collaborations, particularly voluntary collaborations.

Nor is it clear how successful this particular type of grassroots collaboration has been in achieving environmental outcomes that actually improve environmental conditions. Are there unique factors for success in grassroots-organized and -managed collaborations? How do participating organizations bridge sectoral differences to achieve goals? Can grassroots collaborations achieve measurable outcomes? A premise underlying this research is that for the foreseeable future progress on environmental problems will include more nonstate interests in defining problems and developing solutions, organized and led by grassroots environmental collaborations constituted to identify, define, and mitigate environmental problems. This book examines these questions with the intent to achieve better understanding of the driving forces involved in the process of founding, organizing, and managing grassroots environmental collaboratives. Answers will help others to be more effective in future collaborations, within the Chesapeake Bay estuary where these cases are examined, in watersheds throughout the United States, and in voluntary, grassroots-led environmental collaborations organized to address other problems.

What Is Collaboration?

The term "collaboration" has been found in the academic literature for several decades. There have been hundreds, perhaps thousands, of articles written on collaboration, across nearly all fields and areas in the social sciences. Though much effort has been devoted to understanding how collaboration happens, somewhat less effort has gone into developing a common definition of the term.

In her seminal work on the topic, Barbara Gray (1989, 5) provides the following working definition of collaboration:

> a process through which parties who see different aspects of a problem can constructively explore their differences and search for solutions that go beyond their own limited vision of what is possible.

This statement provides a starting point through which we can begin to better specify the meaning of the term. Madeleine McNamara (2012, 391) builds on this definition, but adds some additional elements:

> Collaboration is defined as an interaction between participants who work together to pursue complex goals based on shared interests and a

> collective responsibility for interconnected tasks which cannot be accomplished individually.

Ann Marie Thomson (2001) adds still other components to collaboration:

> Collaboration is a process in which autonomous actors interact through formal and informal negotiation, jointly creating rules and structures governing their relationships and ways to act or decide on the issues that brought them together; it is a process involving shared norms and mutually beneficial interactions.

While these three definitions have certain elements in common (and, indeed, the latter two definitions were influenced by Gray's definition), they all add additional, different elements. Does collaboration include all of these items, or only the common elements?

During a graduate seminar in collaboration taught at Old Dominion University in the spring of 2012, the seminar participants went through an exercise in which they made note of the definitional elements of collaboration found in a subset of ten journal articles assigned for the course. The object of the exercise was to illustrate the lack of definitional agreement in the literature. Collectively, the students listed nearly sixty component elements of collaboration found in this small set of articles. When expanded to a longer list of about thirty articles from the syllabus, the list grew to nearly one hundred elements.

From the longer list, the students identified a list of the nine most commonly cited elements of collaboration. All of these elements were mentioned in at least three articles; there were an additional fourteen elements listed in two or fewer articles:

- *Everyone brings assets to the table.* Participants in a collaboration all contribute resources to the joint effort. They do not necessarily bring assets of equal value, but they must contribute something.
- *Trust.* Because collaboration is voluntary, participants must be convinced that other participants are acting in good faith, and are trustworthy partners in the collaborative effort. Trust is also a central component of social capital (see Putnam 2000).
- *Common goals.* Collaboration works when the participants have a common goal in mind. If the goals of participants are different, the collaborative effort will fail.
- *Assumption of shared risk.* Working with others is an inherently risky business, and a failed collaboration can mean a loss of time, resources, and relationships; it can also mean that an important goal goes unrealized. Risk must be distributed among all participants, and not fall on just one or two members of the collaboration.
- *Voluntary participation.* Participants in a collaborative effort are there by choice. Coercion is very different from collaboration.[1]

- *Mutual benefit.* Voluntary participation means that all participants expect to receive some positive benefit from their efforts. Solutions that reward some at the cost of benefits to others are unlikely to succeed; all participants must feel as though they gain some clear benefit from participation.
- *Interdependence.* Because collaboration involves the creation of a new kind of organizational structure, the participants in the structure exhibit high levels of interdependence—they cannot accomplish the tasks of the group without the contributions of their fellow participants.
- *Flat, not hierarchical, organization.* While traditional bureaucratic organizations have positions that are arranged hierarchically—that is, in levels above and below each other—collaboration is a very flat organizational form. In a very real sense, all participants are equal in stature within the organization, and formal authority arrangements are not found in these structures.[2]
- *Social capital.* At the heart of collaboration is a willingness to work with others in a community in order to accomplish a common goal. Inherent in the idea of "community" is a willingness and ability to work with other members of the community.

Taken together, these elements begin to provide a working framework for a definition of collaboration and, ultimately, a framework through which collaboration may be viewed. Chapter 2 develops such a framework; that framework in turn becomes the "lens" through which we view and analyze the activities of our three case studies.

Social Capital as Motivation for Collaboration

Robert Putnam (2000, 19) defines social capital as

> connections among individuals—social networks and the norms of reciprocity and trustworthiness that arise from them. In that sense social capital is closely related to what some have called "civic virtue." The difference is that "social capital" calls attention to the fact that civic virtue is most powerful when embedded in a dense network of reciprocal social relations.

Social capital, and thus collaboration, is dependent on networks of individuals willing to work together to achieve a goal. While goals for communities may be very broad and loosely defined ("we want to live in a safe community"), goals in collaborative structures are generally defined more narrowly. Social capital is a broader force in a community, while collaborations (as a form of organization) are created for a specific purpose ("we want to make the river fishable and swimmable by 2015").

Since collaboration is dependent on networks of individuals, it is imperative that social capital norms of trust, reciprocity, and efficacy are developed as quickly as possible to support and sustain the collaborative effort. The challenge for each group is to translate the social capital existing within the group to a much larger collection of diverse and often competitive stakeholders. Pierre Bourdieu (1985), one of the first to develop the concept of social capital, offered the insight that social capital of any significance can seldom be acquired without investing some material capital.

As we will see, the founders of two of our case study organizations brought material resources that could be leveraged in a collaborative setting. The challenge for our third case was to enlist participation and develop social capital without the tangible material capital Bourdieu (1985) considered a necessary requisite to social capital accumulation. Neither were there any models in the early 1990s for what they planned. Their success depended on their ability to enlist as participants large industrial firms typically not allied with environmental values, government regulators, scientists, and nonprofit organizations and to build bonds of trust, reciprocity, and efficacy as quickly as possible within that larger organization. It is important as well to keep in mind that all three groups built these social bonds within the political cultures of their respective watersheds. Though we argue in each case study that most of the cities in the watersheds are either a mix of individualistic-traditionalistic or moralistic-traditionalistic political cultures, they were until recent decades traditionalistic according to Daniel Elazar (1984).[3] Traditionalistic cultures do not value participation, nor provide much public space for political efficacy to develop among the general population. Even though the political cultures in this study are migrating away from the purely traditionalistic, the three collaboratives operated in an environment historically resistant to change and civic engagement, thus making very real contributions by both broadening and deepening the values of participatory democracy among the institutional stakeholders and the general population. The dynamic process of social capital accumulation and its role in motivating further participation by stakeholders is central to this study.

Nature of the Problem

Environmental degradation is a "wicked problem." Wicked problems (Rittel and Webber 1973) are so large, and so complex, that successful action defies the use of common approaches and solutions. In the case of water quality, the degradation can stem from either point sources (pollution emanating from a clearly definable, geographically specific location, such as the discharge from a wastewater treatment plant) or nonpoint sources (pollution emanating from a large, geographically dispersed area, or from many small sources, such as fertilizer runoff from suburban lawns). While both the national and state governments address point source pollution through regulatory frameworks and a permitting process, efforts to address nonpoint pollution have proved much more challenging. Moreover, permitting regimes and regulatory frameworks can prevent current and future pollution from point sources, but they are not effective mechanisms to address historically polluted waters—creosote dumped in a river fifty years ago, for example.

Indeed, legacy pollution is an exceptionally wicked problem for several reasons. First, regulatory frameworks for most industrial pollution are a relatively recent phenomenon. If a toxic substance found in a river was dumped there at a time when it was legal to do so, it makes little sense to hold a company liable for an action that was legal at the time it happened. Second, even if we were to decide that a company is liable for its past actions, the company would have to be in business in order to hold it liable for past actions. While large industrial chemical companies may still be in business, smaller companies tend to go out of business or change locations. Third, determining the specific source of some forms of pollution is exceptionally difficult. If two adjacent companies use polychlorinated biphenyls (PCBs) in their plants and PCBs are found in the river, how do we determine which company dumped the waste in the river?

Finally, remediation efforts are often prohibitively expensive. In 1980, for example, Congress enacted the Comprehensive Environmental Response, Compensation, and Liability Act (CERCLA; more commonly known as Superfund), legislation that was specifically intended to provide resources to prosecute polluters, and to provide for remediation efforts when a responsible party could not be identified or held liable.[4] Initially funded at $1.6 billion, the Superfund program has endured years of budget shortfalls, leaving the program significantly underfunded (see Rosenbaum 2002; see also Vig and Kraft 2000). As of this writing, the US Environmental Protection Agency (EPA) lists approximately 1,313 Superfund sites on the National Priority List (US Environmental Protection Agency 2012a), with estimated remediation costs ranging from $75 billion to more than $120 billion (Congressional Budget Office 1994).[5]

On a more limited scale, the costs to address pollution in the Chesapeake Bay are also significant. In January 2003, the Chesapeake Bay Commission released a report that calculated the costs for Bay cleanup between Fiscal Year 2003 and Fiscal Year 2010 at $18.7 billion, with only $12.8 billion in revenue identified over that period—a $5.9 billion shortfall based on estimates (Chesapeake Bay Commission 2003, 3). A recent study by the Hampton Roads Planning District Commission (2011) estimated the cost of meeting the stormwater requirements of the new Bay Total Maximum Daily Load (TMDL) requirements at more than $9.7 billion just for the Hampton Roads area. With pressures to reduce government spending at the federal, state, and local levels, it seems clear that the resources necessary to address these environmental needs will not be forthcoming from government.

Collaboration provides an attractive option for these cases, in that collaborative structures can bring both additional resources and new approaches to bear on a problem. As Gray (1989, 5), suggests, "[c]ollaboration is based on the old adages that 'two heads are better than one' and that one by itself is simply not good enough!" If collaboration is rightly thought of as an alternative form of organization, then the purpose of the collaboration can be thought of as a means to increase organizational capacity without the need for an organization (or a political entity) to commit additional resources to a problem. At the citizen level, collaboration is a means to gather a critical mass of resources that individual citizens lack. At the organizational and individual levels, then, wicked problems provide an externally driven incentive for collaboration.

APPROACH OF THE BOOK

As noted earlier in this chapter, our focus is on the development and operation of grassroots, collaborative environmental organizations; we are particularly interested in the collaboration processes, strategies, and activities that lead to desirable environmental outcomes. We examine these elements by observing the activities and operations of three grassroots environmental organizations located in southeastern Virginia, an area known as Hampton Roads (or by its more historic name, Tidewater Virginia).

Each of the three organizations is structured around a specific tributary of the Chesapeake Bay—the Elizabeth River, the Lynnhaven River, and the Nansemond River. Our inquiry is guided by a comparative case-study approach (Yin 1994), and the cases are analyzed through a lens of collaboration (presented in chapter 2). Our goal for this book is not to develop a grand theory of environmental collaboration, but rather to better understand the elements of collaboration as we find them in these settings. While the framework will guide our inquiry and provide a basis for commonality between the cases, we also offer practical knowledge about "what works" (and, conversely, what doesn't work) in these settings—in short, we present lessons guided by the real-world experiences of these collaborations.

A Challenge to Collaboration

In his highly regarded critique of past efforts to address environmental degradation in the Chesapeake Bay estuary, Howard Ernst (2010) distinguishes between three different approaches to questions of environmental protection. Adherents of the "dark green" approach see access to a clean and healthy environment as a fundamental human right, and that it is the duty of government to employ its legal authority in defense of these rights. The "Dark Greens" thus believe that traditional command-and-control mechanisms, such as environmental regulation, are necessary to achieve desirable outcomes. On the other hand, adherents of the "light green" approach see themselves as stewards of the environment—a healthy environment is desirable, but it is not a fundamental human right. In this model, citizens have a responsibility to identify the source of pollution and "to appeal to the culprits' sense of responsibility to change their harmful behaviors" (Ernst 2010, 6). The "Light Greens" thus rely on tactics such as education, consensus building, moral persuasion, and compromise.

A third group in the mix is what Ernst refers to as the "Cornucopians." This group believes that the free market is a better mechanism to solve common problems, and that there is no need for government interference in a problem that markets can eventually solve. Cornucopians are concerned with the environment as a source of economic inputs, and focus much more on individual property rights, as opposed to collective property rights. Economic growth and resource maximization dominate this line of thought, and Cornucopians eschew more traditional command-and-control solutions.

For Ernst, the Chesapeake estuary has become a political "dead zone" because interest-group politics no longer functions effectively in this policy arena. He argues that the "Light Green" approach of consensus and compromise has effectively marginalized the other groups politically, such that the institutions and protocols developed for estuary protection are grounded in a consensual, voluntary, but largely ineffective, policy regime. By marginalizing the other approaches and thus reducing political conflict, there is no longer a meaningful political process that seeks to address the inherent value conflicts in a constructive manner. We are thus stuck, Ernst argues, in a "feel good" mode in which severe environmental problems go unaddressed, but in which both elected and appointed officials, as well as a multitude of citizen groups, foundations, and scientific organizations can claim "success" (in spite of growing scientific evidence to the contrary).

As persuasive a case as Ernst makes as a "call to arms" for stronger and more effective environmental advocacy, our work in this book differs in some important ways. First, Ernst is focused on the entire watershed, an area of some sixty-four thousand square miles spread across six states and the District of Columbia (Ernst 2003). In comparison, our focus is on just three of the thousands of creeks and rivers that flow into the Chesapeake Bay. Second, while the organizations created to address environmental degradation in these three rivers regularly interact with estuary-wide organizations, our focus is on the individual organizations and the citizens who contribute to these organizations. Third, Ernst focuses on voluntary collaborative efforts between large-scale organizations, an environment in which individual citizens play almost no role. In this setting, organizational actors are asked to participate in a process in which there is little incentive to participate, and even less incentive to pledge resources to the common effort. Moreover, years of failed attempts, coupled with long-standing organizational and intergovernmental rivalries, conspire to make effective agreements (and outcomes) a rarity.

Lastly, and perhaps most importantly, we view consensus and collaboration as essential and desirable outputs of a healthy democratic political process. While we share Ernst's frustration and disappointment over decades of ineffective action and billions of dollars spent without significant environmental improvement, we view the collaborative efforts described in this book as a reaction to (as opposed to further evidence of) the failures of the policy regime in place at the larger political level. Each of the three collaboratives described in this book are products of citizen frustration with the inability of existing efforts to address serious environmental degradation in the waterways in their communities. These groups are not "dark green" or "light green"; they consist largely of pragmatists who share a common goal. They employ collaboration, consensus-building, and engagement precisely because these techniques work. Their objectives are limited, as are their resources, but they have ultimately proved to be capable of achieving measurable gains in environmental quality in their areas of interest. More importantly, the use of collaboration and consensus-building may well be consistent with the dominant political culture (Elazar 1984) in Virginia (McNamara 2010).[6] The fact that they have enjoyed some success utilizing the techniques and approaches dismissed as ineffective by Ernst is not a refutation of Ernst's argument, but rather provides a more nuanced analysis of these patterns at a different level of policy activity. In short, we believe Ernst's dismissal of the positive effects of collaboration is, at best, premature. The connection of citizens to "place" is much more pronounced, and more easily defined, at a local level. While it may be challenging to employ collaboration over a large geographic area (for example, the Chesapeake Bay watershed), collaboration at a local level may be much easier to accomplish. It is hard for people to work with others even fifty miles away, but it is easier to work with neighbors in the same community.

Because *It's My Back Yard!*

Most readers of this book will be familiar with the acronym "NIMBY" (Not In My Back Yard). NIMBY is employed as a means to describe citizen opposition to perceived negative changes to the environment in which they live; indeed, the term itself has a strong negative connotation for most people. As a call to arms, NIMBY has been used to oppose the siting of nuclear plants, prisons, mining operations, large-box retail stores, and a plethora of other proposed projects. Citizens see these projects as being detrimental to property values, quality of life, or the health and welfare of the community.

Gray (1989, 206) contends that the NIMBY problem stems "from an unequal distribution of costs and benefits." A community targeted for a potential project that is perceived to cause harm places significant costs on the residents of the community, and the community objects to bearing the disproportionate costs of the project. Because people who are likely to receive a small benefit from such a project are less likely to offer a position than those who will bear the cost, the predominant voices are those of dissent and objection. The self-interest of those bearing the costs drives the behavior, and may manifest itself either in outright opposition to the project, or an attempt to secure compensation to offset the perceived costs (Gray 1989). The power of the "NIMBY effect" is well documented, as is its reliance on citizen commitment to a sense of community and self-interest.

However, this connection to "place" can also be a powerfully motivating and positive force, for exactly the same reasons it is successful as a negative force. Citizens who wish to protect their immediate surroundings from perceived harm can also be motivated to address environmental issues in their communities because the problems are local, immediate, and salient. The environmental problems may or may not be visible, but the potential for additional harm from inaction (or continued inaction) can motivate people to get involved in efforts to address the problems. In short, people get involved because of a different conceptualization of the same motivation—*Because* It's My Back Yard (BIMBY).

The connection to place upon which BIMBY rests is consistent with the basic elements of social capital. If social capital and civic virtue are a result of "a dense network of reciprocal social relations" (Putnam 2000, 19), then like-minded citizens in those networks will work together to tackle problems common to the community to which they belong; perceived environmental degradation is such a problem. Indeed, high levels of trust within a community might mean that even residents who are less environmentally motivated than their neighbors will participate because of a broader shared vision of a desirable community. In short, it is easier to motivate citizens to engage in something that is immediate to their community and important to their fellow citizens. BIMBY is thus a positive force, in that it becomes a motivation for people to work together for internally driven change, rather than a force to oppose externally driven change.[7]

We expect that the connection to place represented by BIMBY is an important factor in understanding how and why citizen collaborations are formed, and also helps explain why they can make a tangible difference in their watersheds. If we are correct, then BIMBY might also help explain why local grassroots citizen collaborations can succeed where large-scale, agency-based collaborations have not been successful. For a family who owns a house on the Lynnhaven River, the environmental problems on the upper Susquehanna River can seem insignificant and distant, and the causal connection between the pollution from the Susquehanna and the problems on the Lynnhaven can seem farfetched. However, the neighbor on the Lynnhaven who overfertilizes his lawn may have a very visible and direct causal connection to the water quality on the Lynnhaven, and can reduce the quality of life for everyone else on the same river. It seems more reasonable that neighbors are more willing to listen to other neighbors, and thus alter their behavior, than to listen to (or care about) others miles away.

Likewise, a watershed the size of the Chesapeake Bay (the largest estuary in the United States) represents a significant challenge. As detailed elsewhere in this book, the Bay watershed spreads across six states and the District of Columbia, and a significant number of people live within its boundaries. Some of the states contributing significantly to Bay degradation—Pennsylvania, for example—do not even border the Bay and receive little to no economic, cultural, or recreational benefit from it. The water quality problems are myriad and complex, and decades of efforts by well-meaning federal agencies, state agencies, and environmental groups have (at best) slowed the rate of environmental degradation in the Bay. Given the significant range of interests present across all of these groups, it is not surprising that collaboration at the estuary-wide level has not been particularly successful. If the critical components of collaboration include elements such as trust, common goals, and the assumption of shared risk, then trying to induce successful collaboration among the dozens of interested parties is a task of Herculean proportions. In short, it may not marginalize interests, as Ernst suggests, but rather cause a paralysis of interests. All might agree that a common goal is to clean the Bay, but how do we go about it? If we decide that the most important goal is to reduce the amount of nitrogen flowing into the Bay from the Susquehanna River, then Pennsylvania bears the brunt of the effort (and cost). On the other hand, if we decide that the primary problem is bacterial pollution from substandard water treatment plants in the District of Columbia, then the District residents (and the national government) bear the brunt of the effort. If we define the problem as nonpoint runoff from farms in Maryland and Virginia, then farm lobbying groups oppose the effort because of the costs placed on individual farm owners. While there may be agreement on the overarching goal, the prioritization of subgoals (and thus action plans) can be debilitating. A reaction to these vexing problems can be to avoid the more difficult questions, and instead adopt a framework of voluntary action that is more market-like in its flavor. Rather than being compelled to change by the use of regulatory frameworks (themselves anti–free market), a voluntary, collaborative approach allows participants to arrive at a decision for which all can claim credit. The fact that such approaches may be ill-suited to the scale of application is irrelevant; what matters is the ability for all participants to claim "success" because of the existence of an agreement. In a manner of speaking, some interests of all parties are satisfied, even if the problem remains.

Moreover, the larger the scale of the problem, the more difficult it is to represent all interests fairly and accurately in the collaborative process. Farmers in Maryland are different from farmers in Virginia, and are represented by different interest groups at the state level. Likewise, a farmer whose land borders the Chesapeake Bay on the Eastern Shore of Virginia is likely to have a different view of the problem than a farmer whose land borders the upper reaches of the Potomac River, over a hundred miles from the Bay. The same is true of a farmer whose land borders a tributary of a tributary to the Bay—the problem seems very remote and distant. These groups of farmers may be said to belong to a "community of farmers," but it is a "community" only in the broadest sense of the term. They are not physically close to one another; they do not share a common history; and they do not regularly interact. At the local level, however, these limitations are much less pronounced, and may be largely absent altogether. By definition, people in a local community do live close to one another. There is a common history and culture in the community, and that culture provides a common basis for a shared vision. Some level of social capital likely exists, and there is regular social interaction between residents of the community. In short, many of the important elements of collaboration are already present in local communities, making the possibility of successful collaboration more likely at the local level.

As a means to address the environmental problems of a large-scale watershed, can local, citizen-based grassroots collaboration be effective? We believe they can, and our three cases provide evidence of both successful collaboration and enhanced water quality resulting from those efforts. The key is to break down the large-scale problems into more manageable pieces, and to address the pieces one at a time. As one of our interviewees said, "the best way to eat an elephant is one bite at a time" (Crocker 2012). Faced with the task of devouring something as large as an elephant, to attempt to accomplish the task in one "bite" will likely lead to great frustration and, ultimately, failure. However, by cutting the "problem" down to more manageable pieces and eating one portion at a time, eventually the entire beast is consumed. The same is true for water-quality problems. Trying to clean the entire Bay at once, with one framework, is a recipe for failure. On the other hand, many small-scale operations, organized around specific waterways and tributaries, can have the cumulative effect that eludes the large-scale attempt. With some relatively minor exceptions, the water-quality issues in the Bay are due to pollutants emanating from tributary rivers and streams—addressing water-quality issues in the tributaries must necessarily improve water quality in the entire estuary.

The clear benefit of this approach is that, if we are correct that BIMBY is a motivating force for positive change, and it is therefore easier to motivate people to contribute to something positive in their own community, then locally based collaborations are much more likely to have a positive impact on water quality in the Bay, in addition to the more obvious contributions to water-quality improvement in each local setting. In this sense, Ernst misses an important variable in his analysis—scale. The problem is not that collaboration is bad, or ineffective; the problem is that it has been applied at the wrong level. Rather than apply collaboration as the utensil to eat the whole elephant at once, it is the utensil to eat the elephant one bite at a time. By taking advantage of the trust, legitimacy, and efficacy of local communities, collaboration can be an effective approach to ecosystem restoration.

This book will explore an additional theme drawn from Ernst's work. While Ernst concludes that collaboration is ineffective, he sees the appropriate path to be found through command-and-control, top-down, regulatory approaches to water quality. Decades of effort to create regulatory frameworks, detailed in chapter 2, have attempted to implement such an approach. More recently, the TMDL requirements implemented for the Chesapeake Bay and its tributaries have attempted to redefine the basis for the command-and-control approach, and move some responsibility for compliance with the requirements from the state level to local governments. Rather than a single set of pollution limits applied across an entire watershed, the TMDL framework relies on scientific studies to take into account the ability of a body of water to absorb pollution, the kinds of pollution in the body of water, and the sources of that pollution. In addition to point source pollution (pollution emanating from a specific known location, such as the outflow from a sewer discharge), the Bay TMDL also requires nonpoint source pollution to be addressed (pollution that does not emanate from a single point, such as fertilizer runoff from farms along a river, or stormwater runoff from streets and parking lots). The Bay TMDL standards require watershed states and local governments to develop "watershed implementation plans" to address these pollution sources, and to implement a series of "best management practices" (BMPs) designed to meet the mandated limits.

Like any top-down regulatory framework, particularly in our current political environment, there is ready-made opposition to the imposition of such requirements. This is another area in which local grassroots environmental collaborations may prove an effective bridge between local governments and citizens. A local collaborative group that understands the purpose of TMDLs can encourage its members (and members can encourage their fellow citizens) to adopt BMPs that can help meet the requirements, but they do so as a nongovernmental entity. Many surveys and studies suggest that citizens are much more likely to trust their fellow citizens than they are their government, so a local collaborative that works with government to encourage citizen use of BMPs can effectively serve to bridge the gulf between top-down regulation and local compliance. By making the case that such practices are good for the quality of the local environment, these groups have a level of trust and efficacy that estuary-wide groups (and, increasingly, governments) lack. Citizen environmental collaborations can thus serve to enhance significantly the effectiveness of top-down regulatory frameworks by encouraging citizen buy-in and compliance in a way that nonlocal groups cannot.

NOTES

1. There is evidence that some agencies are mandated by law to collaborate with other agencies in order to address identified problems. It is unclear whether such "mandated collaboration" is truly collaboration, or whether it is better thought of as "coordination" or "cooperation."

2. While we realize that informal authority, driven by differences in age, rank, expertise, resources, or other factors can and do exist, these differences are, by definition, not formalized in a collaborative setting.

3. The definitions of the different types of culture are presented in chapter 2, and the political cultures of the respective watersheds are discussed in each case-study chapter.

4. CERCLA was created in response to issues of the historical dumping of toxic waste. In the 1970s, the discovery of large amounts of toxic waste at places like Love Canal and Times Beach spawned public outrage, and spurred Congress to create the Superfund program. In both the Love Canal and Times Beach cases, the waste had been in place for decades, and no responsible party existed from which to secure remediation funding. See Mazmanian and Morell (1992) for a more complete analysis of CERCLA.

5. While toxic waste accounts for only a small portion of the pollutants in the Chesapeake Bay watershed, it is worth noting here that there are currently seventeen Superfund sites listed in the Elizabeth River watershed, and another two sites in the Nansemond River watershed.

6. McNamara (2010) makes a persuasive case that traditionalistic cultures may indeed be fertile ground for collaboration. This issue is discussed more fully in chapter 2.

7. We acknowledge that the similarities in motivation between NIMBY and BIMBY are strong, and that NIMBY also relies on (and can help build) social capital. The important difference is the qualitative difference in motivation between the two. NIMBY represents a negative force (prevent an undesirable change from occurring); BIMBY represents a positive force (effect a desirable change from the present condition). In Gray's (1989) terms, BIMBY occurs when a community perceives a net collective benefit that is greater than the current condition. In this case, the costs of action are concentrated within the community, but the benefits are also perceived to be concentrated. Benefits may also spill over into other communities, but since the benefits of environmental action are most immediately felt in the local community, residents experience a direct positive payoff of their investment.

TWO

A Framework for Collaboration

Building Social Capital, Improving Environmental Outcomes, and Maximizing Democratic Values

Collaboration is based on the simple adages that "two heads are better than one" and that one by itself is simply not good enough!

—Barbara Gray (1985, 5)

The proliferation of environmental grassroots collaborations over the past twenty years signals a shift in American governance with regard to environmental policy and natural-resource management (Lubell et al. 2002). This is evident in watershed partnerships, where the governance of watershed management has predominantly shifted from a traditional centralized approach to reflect decentralized, comanaged arrangements made up of multisector stakeholders coordinating their resources and activities across vertical and horizontal open systems within a collaborative setting (Genskow and Born 2006; Gerlak and Heikkila 2007; Imperial 2005). Watershed collaborations, or partnerships, operate along a spectrum of public and private arrangements, in which the activities of stakeholders range from highly formal processes coordinated among governmental actors to informal arrangements organized and managed by local citizens' groups (Hardy and Koontz 2009; Koontz and Johnson 2004; Leach and Pelkey 2001; Margerum 2008). Watersheds typically span across multiple jurisdictions as well as cross economic and social boundaries (Kenney et al. 2000; McGinnis 1999; McKinney and Harmon 2004). Consequently, these collaborative arrangements pose unique challenges with regard to environmental protection and managing common-pool resources.

The increase in watershed partnership formations throughout the United States is changing the landscape of democratic governance. Government and citizens are working together and actively engaging in finding solutions to intractable environmental problems (Kettl 2002; Sirianni 2009). Since their development, these partnerships have captured the attention of researchers and scholars from multidisciplinary fields such as ecology, environmental law, sociology, political science, and public administration (Goldfarb 1994; Kenney 2000; Leach and Pelkey 2001; Sirianni 2009; Weber 2000). This chapter develops a framework for analysis and comparison of the three case studies examined in this book, and reviews the extant literature on watershed collaborations. We employ these scholarly works to glean insight on what conditions create these partnerships, how they operate, and why they continue to proliferate. We conclude the chapter with a brief discussion of the methods employed in this study.

A MODEL OF WATERSHED COLLABORATION

As noted in the preceding discussion, our model draws heavily from David Easton's (1965) systems model.[1] Other similar models may be found in the extant literature; perhaps one of the better-developed examples is that offered by Paul Sabatier, Christopher Weible, and Jared Ficker (2005) in their study of watershed management. Like the present work, the model offered by Sabatier, Weible, and Ficker is loosely based on the systems model, but the models differ in some important ways. First, the focus of the work by Sabatier, Weible, and Ficker is on watershed management, where our focus is on watershed restoration. Second, while our attention is specifically focused on citizen-centered grassroots efforts, Sabatier, Weible, and Ficker include collaboration between large institutional actors. While there are institutional actors included in the collaborations under study in this book, the focal points differ between the two works. Third, our work is specifically concerned with the role of social capital in the formation and operation of these collaborative structures. We treat social capital as an antecedent, a process variable, and an output of the collaborative process.

We also find inspiration in the work of Elinor Ostrom (1990; 1986) and her model of institutional analysis. Indeed, her model of institutional analysis bears a striking resemblance, in the abstract, to Easton's (1971; 1965) systems model. Her concepts of "states of the world," "action arena," and "results" (Ostrom 1986) are similar in definition to inputs, process, and outputs of Easton's model. Likewise, her concept of "community" can be conceived of as a combination of Easton's "demands" and "support" components of the "inputs" box.

Figure 2.1 depicts the model developed for our study. In the abstract, our model consists of a set of preconditions (contextual variables) that define the setting for collaborative action. The collaboration process includes a set of variables that, collectively, describe the actions of the participants in watershed activity. As a result of the collaboration process, three kinds of results are produced—changes in social capital, short-term changes in the watershed (outputs), and long-term changes in water quality (outcomes).[2] All three "results" categories are linked to the contextual variables through a set of feedback loops. In other words, the model includes a dynamic component in which previous actions inform current actions, which in turn inform future actions. As collaborative efforts result in the development of a watershed action plan, for example, participants gain additional trust in their fellow participants, which in turn enhances the level of social capital in the community. As nonparticipants observe the initial successes of the group, they may be encouraged to join the group, and trust builds concomitantly. Likewise, as water quality in the watershed improves, the context in which the group operates changes, and the focus of the group can change to address a different set of issues.

This framework also provides a structure for the case studies in chapters 4, 5, and 6. We have organized each case study according to the elements of the model; they are presented in the same order in each

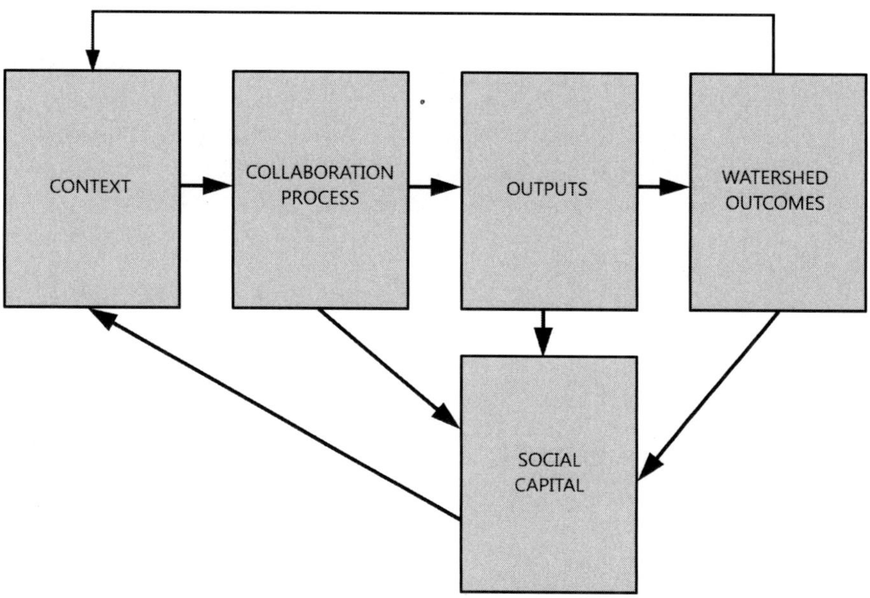

Figure 2.1. A Model of Collaboration

chapter to facilitate comparison across the cases. The following sections of this chapter draw on the literature on watershed collaboration to help identify and define each of our major conceptual elements, and to specify the variables employed to represent each conceptual element. We begin with an examination of the contextual elements of the model.

THE CONTEXT OF WATERSHED COLLABORATION[3]

Grassroots environmental collaboration takes place in a context that is definable at two important levels. The first is a national/international level, and includes the set of national laws and policy initiatives, and interest group activities that define national politics (see Kingdon 1984). It can also include international treaties, pollution originating in other parts of the world, and other large-scale factors. Conversely, the local context is located within this larger setting, and includes factors that are specific to the local community. These factors include political culture, nature of the problem in the watershed, resources, and social capital. All of these factors impact the setting in which grassroots collaboration occurs.

The Broad Context

The literature identifies several overarching themes that are attributed to the development of environmental collaboration, including policy shifts (Gerlak 2005; Kraft and Scheberle 1998; Sabatier, Weible, and Ficker 2005), institutional and market failures (Bryson, Crosby, and Stone 2006; Gray 1985; Weber 2000), public demands (Cooper, Bryer, and Meek 2006; McKinney and Harmon 2004; Nelson 1999), scarcity of natural resources (Gerlak 2005; Gerlak and Heikkila 2007; Kettl 2002; McKinney and Harmon 2004), and problem severity (Cline and Collins 2002; Gray 1985; Lubell et al. 2002). These themes form the basis of the broader context for grassroots environmental activity.

The watershed literature identifies changes in the role of the federal government and policy agendas as factors leading to watershed collaboration (Gerlak 2005; Kraft and Scheberle 1998; Lubell et al. 2002; Nelson 1999; Sabatier, Weible, and Ficker 2005). A number of authors examine the development of watershed partnerships from a historical perspective (Gerlak 2005; Kraft and Scheberle 1998; McKinney and Harmon, 2004; Sabatier, Weible, and Ficker 2005; Weber 2000). In this regard, historical perspectives provide reseachers and scholars a contextual framework in which to view the evolution of America's water policy.

Prior to the nineteenth century, management of natural resources was predominantly under the purview of state and local governments (see McKinney and Harmon 2004; Gerlak 2005). During this time period, decisions over the usage and allocation of water resources were left in the hands of property owners and private industry. Political agendas regarding natural resources focused on commerce and economic development (Gerlak 2005; Sabatier, Weible, and Ficker 2005). As a result, the welfare of communities and the management of natural resources were at the discretion of the marketplace (McKinney and Harmon 2004; Sabatier, Weible, and Ficker 2005). Furthermore, in western states the doctrine of "first in time, first in rights" governed the use of water and natural resources (Gerlak 2005; McKinney and Harmon 2004; Sabatier, Weible, and Ficker 2005). This doctrine suggested that "the first person to put water to use gained an exclusive, preemptive right to that water" (McKinney and Harmon 2004, 34).

The push to develop the West spurred governmental policies that encouraged allocation of public resources (i.e., water, land, and minerals) to western settlers (McKinney and Harmon 2004). These policies were predicated on the notion that entrepreneurs acting in their self-interests ultimately benefitted the public good. However, these policies created devastating environmental consequences, such as land erosion from the commercial harvesting of timber, which in turn led to devastating downstream flooding as well as severe overgrazing by ranchers (McKinney and Harmon 2004). Ultimately, these practices resulted in demands from conservation and preservation groups for governmental intervention (McKinney and Harmon 2004; Sabatier, Weible, and Ficker 2005).

The opportunistic practices of private actors, coupled with the abusive waste of natural resources, resulted in new federal reforms aimed at conserving and protecting common-pool resources. These reforms include the advent of a national park system, beginning with Yellowstone in 1872 (McKinney and Harmon 2004). The passage of the General Revision Act of 1891 gave the president the authority to protect public land and forests, subsequently leading to the creation of the US Forest Service in 1905 (Culhane 1981), and the Reclamation Act of 1902, which appropriated federal funding for the construction of irrigation infrastructures (Gerlak 2005; McKinney and Harmon 2004). These initatives signaled a pivot in the national agenda with regard to natural-resource management and environmental values.

The concept of watershed and establishing regional boundaries as a unit of governance emerged during the Progressive Era (Gerlak 2005; Sabatier, Weible, and Ficker 2005). Progressives argued that market failures required governmental policies that provided stricter controls on the uses of natural resources (Sabatier, Weible, and Ficker 2005; Weber 2000). Legislation enacted during this time period expanded federal authority over watershed management. For example, the River and Harbors Act of 1925 authorized the US Army Corps of Engineers (USACE) to develop comprehensive basinwide strategies for "irrigation, navigation, power production and flood control" (Gerlak 2005, 235). Environmental progressive agendas continued during the Great Depression (Sabatier, Weible, and Ficker 2005), and New Deal initiatives enacted during the Franklin Roosevelt administration created social policies aimed at improving the plight of Americans (Gerlak 2005; Goldfarb 1994; Sabatier, Weible, and Ficker 2005). These policies included creating public-works programs to stimulate jobs and regional development through the construction of public hydroelectric power facilities in depressed regions (Goldfarb 1994; Nelson 1999; Sabatier, Weible, and Ficker 2005). Among the first of these public projects was the Tennessee Valley Authority (TVA), incorporated in 1933 (Gerlak 2005; Goldfarb 1994; Nelson 1999; Sabatier, Weible, and Ficker 2005). The TVA created a "multipurpose" policy strategy for addressing water management and economic development (Nelson 1999). These projects expanded the scope of the federal government with regard to water management (Gerlak 2005; Goldfarb 1994; Nelson 1999; Sabatier, Weible, and Ficker 2005), yet water-quality controls continued to fall under the purview of state and local governments (Gerlak 2005; Sabatier, Weible, and Ficker 2005).

The period from 1940 to 1965 is characterized by activity in what Henry Caulfield (1989) refers to as the latter years of the conservation movement, punctuated by a shift from an industrial to a postindustrial society in the United States (Kraft and Vig 1994). The postwar years were noted for a dearth of environmental legislation; while there were many proposals for increased setasides for public lands and national parks, there was little movement toward meaningful policy outputs. The election of John F. Kennedy in 1960 brought increased attention to the environment and conservation, and the publication of Rachel Carson's *Silent Spring* in 1962 served to increase public awareness of the environment.

While Kennedy did not live long enough to see his efforts at environmental protection bear fruit, Congress passed the Land and Water Conservation Fund Act and the Wilderness Act into law in 1964. These two pieces of legislation formed the nucleus of a new and developing environmental movement in the United States. A new focus on watersheds as environmental systems also began to take hold, and areas such as Cape Cod, the Delaware Water Gap, and the Colorado River began to benefit from the increased focus of these efforts (Caulfield 1989). During the latter part of the 1960s, these efforts were further boosted by new amendments to the Water Pollution Control Act, the Endangered Species Act, and the Water Quality Act of 1965.[4] All of this activity ultimately set the stage for the formation of the Environmental Protection Agency (EPA) and the passage of sweeping federal environmental legislation organized around pollution media.

The late 1960s ushered in the environmental movement (Gerlak 2005; Goldfarb 1994; Sabatier, Weible, and Ficker 2005). Decades of federal initiatives aimed at encouraging economic development created a boom in industrial activity resulting in the visible deterioration of the environment (US Environmental Protection Agency 2012b). As a result, societal values regarding the environment and protection of natural resources began to shift. The literature attributes changes in environmental values to four factors: (1) technological advancements in ecosystem science that enhanced the capacity of detecting environmental issues; (2) a rise in socioeconomic standards that increased the demand for outdoor recreation; (3) a growing distrust in government's ability to balance economic interests with environmental interests; and (4) a rise in environmental disasters such as toxic chemical spills and beach closures (Culhane 1981; Goldfarb 1994; Gray 1985, 1989; McKinney and Harmon, 2004; Sabatier, Weible, and Ficker 2005). Consequently, groups from all segments of society, such as environmentalists, academics, research foundations, and citizens, demanded governmental action (Gray 1985, 1989; US Environmental Protection Agency 2012b). Specifically, these groups demanded stronger governmental oversights and regulations on industries (Sabatier, Weible, and Ficker 2005).

The 1970s brought about significant environmental reforms beginning with the creation of the EPA (Sirianni 2009; US Environmental Protection Agency 2012b). The creation of the EPA signaled the federal government's resolve to address environmental issues through centralized governance using command-and-control mechanisms. Subsequently, major environmental protection policies were enacted during this timeframe. These include the National Environmental Policy Act (1970), Water Pollution Control Act (1972), and the Coastal Zone Management Act (1972) (Sabatier, Weible, and Ficker 2005; US Environmental Protection Agency 2012b).

The Water Pollution Control Act imposed stringent water quality regulations on states and localities (Adler, Landman, and Cameron 1993; Heilman and Johnson 1992). In order to meet these standards, federal grants were available to states and local governments in order to implement new water-treatment strategies and pay for upgrades to water facilities (Vig and Kraft 2000). However, federal funding to states drastically decreased during the Reagan administration (Gerlak 2005; Morris 1997, 1996). This shifted the primary costs of these mandates to states and localities, codified in the Water Quality Act of 1987 and the establishment of the Clean Water State Revolving Loan Fund program (Kraft and Scheberle 1998; Morris 1997; Travis, Morris, and Morris 2004). These changes strained intergovernmental relations between states and the federal government (Kraft and Scheberle 1998; McKinney and Harmon 2004; Nelson 1999; Sabatier, Weible, and Ficker 2005). States and localities criticized the "inflexibility and inefficiency of 'one-size-fits-all' regulations," arguing that federal mandates failed to consider local economic impacts or address local environmental interests (Nelson 1999, 355).

While command-and-control regulations successfully reduced pollution from identified (point) sources such as factories and sewage-treatment plants, they proved ineffective at addressing nonpoint source pollution (Kettl 2006; Lubell et al. 2002). According to the EPA (2010a, para. 4), "[S]tates report that nonpoint source pollution is the leading remaining cause of water quality problems." Nonpoint source pollution is derived from a combination of groundwater runoff and man-made pollutants that flow into rivers, lakes, and streams that ultimately feed into estuaries (US Environmental Protection Agency 2010a). Furthermore, nonpoint source pollution crosses jursidictional boundaries and is the culmination of natural and collective human activities (Kettl 2006; Lubell et al. 2002). Policy scholars point out that the pervasiveness and problem severity of nonpoint source pollution requires new policy tools that supplant regulatory mechanisms with a holistic approach to water-resource management (Heikkila and Gerlak 2005; Nelson and Weschler 2001; Smith and Ingram 2002; Weber 2000).

The genesis of watershed partnerships represents a shift within the political and social landscape. Scholars attribute the reinventing government movement begun during the Clinton administration as the impetus for grassroots environmental collaboration (Weber 2000). In an effort to promote efficiency and effectiveness within federal agencies, the EPA implemented innovative programs such as the Community-Based Environmental Protection program and the Community Action for a Renewed Environment (US Environmental Protection Agency 1995; Sirianni 2009). These programs seek to build capacity through partnering with local stakeholders; community actors collectively engage with state and/or nonstate actors in designing and implementing water-quality solutions that balance environmental and economic interests specific to regions (Goldfarb 1994; US Environmental Protection Agency 1995).

In contrast to traditional command-and-control institutions that delineate authority and control resources, collaborative arrangements diffuse authority across a network of actors who mutually agree to share their resources and coordinate their activities to pursue a common objective (Gray 1989; Wood and Gray 1991). Bryson, Crosby, and Stone (2006) and other scholars emphasize that collaborative arrangements are complex in nature. These organizational structures operate within a dynamic environment. Thus, operating effectively within these structures requires distinctive processes and procedures that facilitate multilevel coordination among a diverse set of stakeholders.

The Local Context

The elements of the local context include political culture, the nature of the environmental problems in the watershed, available resources to address the problems, the history of the region, and the amount of social capital present in the community at the outset of the collaborative efforts.

Political Culture

The concept of political culture dates back to the early work of James Wilson and Edward Banfield (1971; 1964; 1963), who were interested in deciphering why some communities seem to be more "public-regarding" than others. They noted that some communities seem to have high levels of civic and political activism, higher voter turnout rates, and more engaged city governments, while other cities seem to be quite the opposite. Their work led Daniel Elazar (1984) to describe similar state-level characteristics he termed "political culture." Elazar's interest was in describing state-level policy choices in a federal system, and he determined that patterns of migration by different religious groups over time determined the three major categories of political culture—individualistic, moralistic, and traditionalistic.[5] Elazar (1984) arrived at his state-level type based on the individual cultures of regions within the states.

The political culture present in a given region serves as an important variable in explaining the kinds of political action possible in that region. When comparing collaborative activity across regions, researchers must account for these differences in culture, since they are likely to impact the kinds of interactions within the setting. However, it is also possible that political culture can vary within regions, as is the case with our research. Although our three cases are located within twenty-five miles of each other, significant differences in local political culture have an important effect on the nature of the collaboration.

Elazar (1984, 115) suggests that

> [t]he individualistic culture emphasizes the conception of the democratic order as a marketplace. In this view, government is instituted for strictly utilitarian reasons, to handle those functions demanded by the people it is created to serve. A government need not have any direct concern with questions of the good society, insofar as it may be used to advance some common conception of the good society formulated outside of the political arena. . . . Since the individualistic political culture emphasizes the centrality of private concerns, it places a premium on limiting community intervention—whether governmental or nongovernmental—into private activities to the minimum necessary to keep the marketplace in proper working order . . . political life within an individualistic political culture is based on a system of mutual obligations rooted in personal relationships.

Individualistic political cultures are not necessarily averse to the idea of collective citizen action outside of the realm of government, but power structures in individualistic societies are not well-equipped to respond to citizen demands on a collective basis. Politicians remain in office because they are able to respond to the demands of individual citizens, and corruption is often expected in this culture (Elazar 1984, 116). In a system where grassroots environmental groups compete with business interests for favorable policies, the market-oriented business interests are more likely to reign supreme.

In contrast, the moralistic political culture "emphasizes the commonwealth conception as the basis for democratic government" (Elazar 1984, 117). A moralistic culture places a great deal of emphasis on notions of the common good, and the purpose of government (and thus public policy) in this culture is to maximize the common good. Moreover,

> In the moralistic political culture, individualism is tempered by a general commitment to communal—preferably nongovernmental, but governmental if necessary—power to intervene into the sphere of private activities when it is considered necessary to do so for the public good or the well-being of the community. (Elazar 1984, 117)

In the moralistic culture, citizens have a moral obligation to participate in public affairs; politics is not limited to the realm of professional politicians. By taking an active interest in their community, citizens work to promote the general welfare, and the view that politics exists for personal economic gain is largely rejected (Elazar 1984, 117). In this environment, we would expect to see citizen groups and government working closely together on efforts that are designed to enhance the general welfare (or quality of life) in the community. In other words, government would be very supportive of grassroots environmental organizations, and would actively support such groups.

Traditionalistic political cultures are marked by "a paternalistic and elitist conception of the commonwealth" (Elazar 1984, 118). This culture sees the role of government as positive, but defines "positive" as "the continued maintenance of the existing social order" (Elazar 1984, 118). In other words, traditionalistic cultures rely on government to prevent social and political change, and societal authority is vested in a limited, self-referential, and self-perpetuating group of elites within the community. Government is not necessarily oppressive, as much as it is disinterested. Elazar (1984, 119) also notes that many traditionalistic communities exhibit "the organic characteristics of the pre-industrial social order." Traditionalistic cultures are generally found in the southern states (such as Virginia), but pockets of traditionalistic culture can be found throughout the nation. In terms of citizen-based collaborative efforts, traditionalistic communities are likely to be disinterested and largely unsupportive of grassroots citizen efforts; to acknowledge such initiatives would be a tacit abdication of authority from the small group of elites to a broader group of citizens. Governments in this culture may not work overtly to quash collaborative efforts, but they will feel threatened by a potential change to the existing social order.

The Nature of the Problem in the Watershed

The nature of the problem in the watershed is another important element of the local context. The perceived problems present, and thus the nature of the collaborative response to the identified problem, can determine the membership of the collaborative, as well as the goals, strategies, and operating principles. If a problem is caused by years of industrial dumping, for example, the kinds of actions possible in the watershed, and the ways in which those actions will be addressed, might be very different from a watershed in which the primary problem is suburban or agricultural nonpoint pollution.

Local communities can also differ significantly in the kinds and amounts of resources available to address pollution issues. In wealthy communities, for example, people have both the disposable income and the free time to contribute to grassroots environmental efforts (see Parisi et al. 2004). Resources might also include the local availability of technical and scientific expertise that can be brought into the collaborative structure. Other factors, such as education levels, perceived quality of life, and other similar variables have been shown to impact the local context (Parisi et al. 2004).

Initial Social Capital

Finally, grassroots environmental collaboration is very difficult to achieve without a sufficient level of initial social capital. While we address social capital in a subsequent section of this chapter, we suggest that there must be a baseline level of social capital available in order to initiate collaborative behavior. In short, citizens must be able to trust their fellow citizens (and, depending on the nature of the collaboration, local institutions) enough to agree to become involved in the initial stages of the efforts. A proposition drawn from this model addresses the growth of social capital as a result of grassroots environmental activity, but we begin from the premise that some level of social capital must exist to launch the initial effort.

COLLABORATION PROCESS

The growing complexities of modern society create the conditions for collaborative arrangements in natural resource management. In this regard, the integrative nature of collaboration changes the dynamics of the process (Gray 1989; Thomson and Perry 2006). Scholarship on environmental collaboration suggests that the efficacy of watershed partnerships requires alternative approaches to organizational processes such as management, planning, and decision making (Campbell, Koontz, and Bonnell

2011; Clark, Burkardt, and King 2005; Gerlak 2005; Leach and Pelkey 2001; Nelson and Weschler 2001; Weber 2000).

In examining the literature on the collaboration process and environmental management, several primary themes emerge, including the role of stakeholders (Bonnell and Koontz 2007; Cigler 1999; Gray 1985, 1989; Genskow and Born 2006; Heikkila and Gerlak 2005; McNamara, Leavitt, and Morris 2010; Leach and Pelkey 2001), resources (Bonnell and Koontz 2007; Moore and Koontz 2003; Smith and Gilden 2002), governance (Bryson, Crosby, and Stone 2006; Koontz and Johnson 2004; Margerum 2008), and institutional activities (Daley 2008; Hardy and Koontz 2009; Schuett, Selin, and Carr 2001). These themes are reflected in the subelements of our model—stakeholders and roles, conveners, resources, rules, and governance structures. We consider these in light of our framework presented earlier in this chapter.

Stakeholders and Resources

Douglas Kenney (2000) emphasizes that collaboration, in and of itself, is not a sufficent incentive to bring stakeholders together. Joseph Bonnell and Tomas Koontz (2007) and Tanya Heikkila and Andrea Gerlak (2005), among others, contend that consensus-based process models present challenges and contraints. Studies on collaborative problem-solving identify transaction costs (i.e., time and money) as one of the primary disincentives that influence the participation of stakeholders in the process (Heikkila and Gerlak 2005; Irvin and Stansbury 2004; Margerum 2007). Collaboration scholars assert that transaction costs can be substantial during the initial phases of the collaboration process (Koontz and Johnson 2004; Kenney 2000; Margerum 2007). In this regard, a critical issue explored by scholars is, what motivates stakeholders to collaborate? The collaboration literature identifies various motivating factors that induce stakeholder participation in the collaboration process, including whether the current state and/or condition serves their interests, and the belief that collaboration will result in a positive outcome (Bryson, Crosby, and Stone 2006; Clark, Burkardt, and King 2005; Gray 1985, 1989; Lubell et al. 2002).

In addition, scholarship identifies conveners, policy entrepreneurs, sponsors, and champions as vital actors in the collaboration process (Cigler 1999; Bryson, Crosby, and Stone 2006; Gray 1985; Heikkila and Gerlak 2005; McNamara, Leavitt, and Morris 2010; Purdy 2012). These actors serve as informal leaders within collaborative settings. Sponsors are participants who identify an issue and recognize that collaborative problem-solving may lead to a better outcome (Policy Consensus Initiative 2012; see also Purdy 2012). Gerlak and Heikkila's (2007, 56) study of four large-scale watershed partnerships identified champions as instrumental in "sustaining the momentum and continued support from various actors for these institutions." In addition, Gerlak and Heikkila (2007) assert that in large-scale watershed partnerships political interests vary; thus effective leadership is essential in developing and maintaining trust among a diverse set of stakeholders.

On the other hand, conveners and policy entrepreneurs are described as influential actors that mobilize key stakeholders to address a problem domain (Bryson, Crosby, and Stone 2006; Gray 1989; Heikkila and Gerlak 2005; McNamara, Leavitt, and Morris 2010). Gray (1985) notes that conveners are partcipants that serve multiple roles such as leader, negotiator, and facilitator. Their legitimacy is based on a variety of factors such as their status in the community (McNamara, Leavitt, and Morris 2010), constituencies (Heikkila and Gerlak 2005), and expertise and/or knowledge in the problem domain (Wood and Gray 1991). Furthermore, Donna Wood and Barbara Gray (1991) assert that in order for conveners to mobilize stakeholders and bring them to the table, several conditions must be present, including: (1) conveners must be viewed as impartial in their approach to solve the problem; (2) they must understand the value in collaborative problem-solving; and (3) conveners must envision the purpose and process of the endeavor.

Heikkila and Gerlak (2005) and Beverly Cigler (1999) found that policy entrepreneurs are instrumental in creating stakeholder buy-in during the initial formation of a collaborative. In this regard, Ann Marie Thomson and James Perry (2006) argue that stakeholder buy-in is an important factor in the collaboration process because participation is voluntary. In addition, John Bryson, Barbara Crosby, and Melissa Stone (2006) assert that leadership positions in collaborative settings are not based on power or authority *per se* but rather the tangible (i.e., financial and human capital) and intangible (knowledge and social capital) resources that a stakeholder brings to the collaboration.

Collaboration scholars argue that assembling the right mix of stakeholders to join a collaborative endeavor is imperative (Gray 1989; McGuire 2006; Wood and Gray 1991). In this regard, the literature suggests that effective conveners identify and select stakeholders based on their ability to create synergies by leveraging resources such as financial, technical, human, and social capital in order to meet their objectives (McGuire 2006). Therefore, linking resources needed to stakeholders' contributions is vital in the collaboration process (Gray 1989; McGuire 2006).

These findings are supported in the watershed literature. The role of conveners is noted in Madeleine McNamara, William Leavitt, and John Morris's (2010) case study of Lynnhaven River NOW, a local community-based watershed group in Virginia. The authors identify the founders of the organization as conveners and credit their "political and financial connections" (McNamara, Leavitt, and Morris 2010, 11) as instrumental in garnering resources and support from key stakeholders. Furthermore,

William Leach and Neil Pelkey (2001) and Allan Curtis, Bruce Schindler, and Angela Wright (2002) discuss the importance of leadership and funding sources in the watershed literature. Leach and Pelkey's (2001) review of thirty-seven watershed partnerships found that twenty-three out of the thirty-seven partnerships surveyed identified funding and effective leadership as the top two factors that impacted organizational goals. Curtis, Schindler, and Wright's (2002) comparative study of two community watershed partnerships, one in Australia and the other in Oregon, revealed that members attributed government funding and leadership attrition to their level of success in community outreach. A consistent finding drawn from this body of work is that leadership and funding are the top two critical factors in watershed collaboration (Curtis, Schindler, and Wright 2002; Leach and Pelkey 2001).

Rules and Governance Structures

Nestled within the collaboration process dimension are the interrelated concepts of governance structures and rules. Viewing collaboration through the lens of public administration, Kirk Emerson, Tina Nabatchi, and Stephen Balogh (2011, 2) define collaborative governance as:

> the processes and structures of public policy decision making and management that engage people constructively across the boundaries of public agencies, levels of government, and/or the public, private and civic spheres in order to carry out a public purpose that could not otherwise be accomplished.

Scholars identify institutional activities as instrumental to collaborative governance because they define the internal operations, functions, and objectives of partnerships (Hardy 2010; Hardy and Koontz 2009). Institutional activities include the development of governance mechanisms such as institutional rules. The literature suggests that governance mechanisms are particularly essential in collaborative settings where the level of automomy among stakeholders is high (Bryson, Crosby, and Stone 2006). Therefore, in the absence of clear lines of authority, rules provide stakeholders in self-governing arrangements a clear understanding of their roles, responsiblities, and how collective resources are utilized (Bryson, Crosby, and Stone 2006; Hardy and Koontz 2009; Leach 2006). Furthermore, scholarship suggests that institutional rules create transparency in the process, thus facilitating trust and mutual understanding among partners (Emerson, Nabatchi, and Balogh 2011; Leach 2006). Leach (2006) asserts that mutual understanding of the process allows stakeholders to determine whether their participation serves their interests. Moreover, William Leach (2006, 103) argues, "To collaborate under false premises is unfair and unlikely to produce sustainable benefits for any of the participants."

Recently, a growing body of watershed scholarship focuses on the unique characteristics among watershed parnterships, including stakeholder composition (Moore and Koontz 2003), group membership (Hardy 2010), institutional level activities (Margerum 2008), and organizational affiliation (Bidwell and Ryan 2006). Findings from this research stream show that institutional structures and interorganizational dynamics play a role in governance, institutional rules, and access to resources as well as in environmental objectives (Bonnell and Koontz 2007; Hardy and Koontz 2009; Koontz and Johnson 2004).

Elizabeth Moore and Tomas Koontz (2003) found that the degree of participation among governmental and nongovernmental stakeholders varies among watershed partnerships. Their study of sixty-four watershed partnerships in Ohio identified three broad types of stakeholder compositions; agency-based, mixed, and citizen-based. Expanding on Moore and Koontz's typology, Scott Hardy and Tomas Koontz (2009) examined levels of institutional activities across the different types of watershed partnerships.[6] Building on institutional levels, Richard Margerum's (2008) analysis of thirty-six watershed partnerships found that coordination activities fell along a continuum of policy, organizational, and action levels. Ryan Bidwell and Clare Ryan's (2006) study of twenty-nine Oregon watershed partnerships revealed that the degree of membership heterogeneity directly influenced their outputs and outcomes. Scott Hardy's (2010) case study of six partnerships operating in the Lake Erie watershed found that the degree of government participation influences resources, issue framing, internal structure, and outcomes.

It is important to note that while there are distinctions among all watershed partnerships, each operates under the core principles of collaboration: participation is voluntary; decision making is consensus based through deliberative dialogue; resources are mutually shared; and partners share a common mission (Bidwell and Ryan 2006; Clark, Burkardt, and King 2005; Shaw 2003; Thomson and Perry 2006). Using Moore and Koontz's (2003) typology as a construct, the following section unpacks the prevailing themes found in the literature on governance and institutional mechanisms.

Agency-based Partnerships

Agency-based partnerships are primarily comprised of institutional actors operating in the public sphere, such as federal, state, local agencies; elected officials; scientific communities; and research universities (Heikkila and Gerlak 2005; Margerum 2008). Heikkila and Gerlak's (2005) research identifies several examples of agency-based partnerships: the Northwest Power and Conservation Council's Fish and Wildlife Program, the Chesapeake Bay Program, the CALFED Bay-Delta Program, and the Comprehensive Everglades Restoration Plan. Robinson et al.

(2011) and other scholars argue that agency-affiliated partners must balance the mission of their respective organizations with that of the collaborative, thus creating a "two-table" dynamic.

In agency-based arrangements, governance of the watershed operates at the regional level (Heikkila and Gerlak 2005). Furthermore, issue framing is broad in scope and crosscuts through multiple geographic regions that impact large populations (Margerum 2008; Robinson et al. 2011). Margerum (2008) points out that institutional-level activities in these partnerships are policy oriented. In this regard, Margerum (2008) contends that group deliberations center on environmental legislation, policies, and water-quality standards. The role of government in these organizations is that of "policy making partner" (Margerum 2008, 489).

Hardy (2010, 594) finds that agency-affiliated groups receive the majority of their resources from governmental entities, grant programs, and "municipal government membership dues." Their access to technical, human, and financial capital allows these partnerships to pursue complex watershed issues on a large scale (Hardy 2010). Robinson et al. (2011) underscore that partnerships in this arrangement rely heavily on governmental support in order to meet their extensive operating budgets; thus their environmental outputs and outcomes may be influenced by political pressures.

Mixed Groups

Tomas Koontz and Elizabeth Johnson (2004) found that mixed and citizen-based partnerships are more prevalent in watershed collaboration than agency-based partnerships. Mixed groups utilize a cross-sector approach to natural resource management (Hardy 2010; Koontz and Johnson 2004). These arrangements are comprised of stakeholders from the public, private, and nonprofit sectors (Koontz and Johnson 2004). Like agency-based partnerships, actors in this arrangement focus on addressing complex watershed issues that cross jurisdictional boundaries and impact large heterogeneous populations (Hardy 2010; Hardy and Koontz 2009; Koontz and Johnson 2004).

Due to the diversity of stakeholders, resources come from a wide range of sources such as "government grants, municipal dues, private donations, and grants from regional charitable foundations" (Hardy 2010, 595). In addition, the watershed literature indicates that mixed groups direct their activities toward on-the-ground projects such as "restoration, land acquisitions, and changes in land use practices" (Koontz and Johnson 2004, 196). Bidwell and Ryan (2006) and others found that in mixed groups, members report a higher level of conflict than in the other two arrangements. Scholars attribute this to the higher level of diversity among stakeholders within these organizations (Bidwell and Ryan 2006; Bonnell and Koontz 2007).

The literature suggests that stakeholder characteristics such as diversity and membership affiliation directly impact the transactional costs of group activities and the formation of watershed partnerships (Bidwell and Ryan 2006; Lubell et al. 2002). Moreover, Koontz and Johnson (2004) note that in highly diverse collaboratives, members devote a substantial amount of time to forging agreements and developing plans. Furthermore, Bonnell and Koontz (2007, 163) argue that these arrangements struggle with the "paradox of scale." Specifically, the task of making broad-scale watershed problems salient to local stakeholders is a challenge (Bonnell and Koontz 2007). In this regard, citizen-based partnerships bridge this gap.

Citizen-based Partnerships

Citizen-based groups are typically linked with a grassroots approach to natural-resource management (Weber 2000). Stakeholders in this group are comprised of an alliance of local community actors seeking to collectively find solutions to local watershed issues (Kenney et al. 2000; Morton 2008; Weber 2000). In addition, Matthew McKinney and William Harmon (2004) argue that community-based collaboratives direct their efforts toward shaping social, economic, and environmental values of a specific locality, thus reinforcing the notion of citizen engagement and democratic deliberative participation. These arrangements are convened "by and for citizens" (McKinney and Harmon 2004, 181). Unlike agency and mixed partnerships, interests among stakeholders are narrow in scope and focus on projects that promote environmental ethics and stewardship (Hardy 2010; Hardy and Koontz 2009). In addition, the majority of actors participating in these collaboratives typically represent themselves rather than an organization (Margerum 2008).

These partnerships operate outside the purview of government agencies. In this regard, government plays the role of a supporting actor through grants, technical assistance, and information concerning water-quality issues in their local watershed area (Clark, Burkardt, and King 2005; Margerum 2008). Margerum (2008) classifies these partnerships as "action collaboratives," whereby stakeholders collectively seek to address local watershed issues through community outreach and public-awareness programs. Their sphere of influence is through their social networks and their common ties to the community (Weber 2009). In addition, Margerum (2008) and others found that "action collaboratives" are more likely to exert pressure on government through advocacy activities such as lobbying and petitioning.

Of the three types, the literature reveals that citizen-based partnerships are the most vulnerable to long-term sustainability threats (Lubell et al. 2002; Lurie and Hibbard 2008). Studies show that access to resources and funding sources are limited in these arrangements (Koontz and Johnson 2004; Lurie and Hibbard 2008). In addition, Hardy (2010) found that these partnerships are highly dependent on voluntary association dues, private donations, and small governmental grants. Consequently, scholars contend that a considerable amount of effort is spent on "chasing the dollar," which diverts activities away from accomplishing organizational objectives (McNamara, Leavitt, and Morris 2010).

Susan Lurie and Michael Hibbard's (2008) review of local community-based collaboratives in Oregon identified several institutional challenges. First, the authors found that increased competition for grant money coupled with the instability of government funding impacts long-term planning. Second, they contend that administrative costs are generally not covered by grants. Consequently, these organizations are dependent on community support and volunteers to donate their time, skills, and facilities to maintain administrative operations (Lurie and Hibbard 2008).

With regard to human and technical capital, Lurie and Hibbard (2008) and other scholars noted that these groups rely on organizational partners for scientific and technical expertise. In addition, Kristin Floress et al. (2009, 1357) identified several factors that hinder the collaboration process in grassroots collaboration, including homogeneous skill sets among group members, ambiguous roles, disagreement among partners over the group's purpose, and "failed problem identification."

A common thread found in the watershed literature is that citizen-based partnerships struggle to maintain adequate financial, human, and technical capital. While this certainly creates challenges in these organizations, scholarship suggests that the strength of these grassroots organizations is their ability to harness and mobilize collective action through the connection of social networks (Floress, Prokopy, and Allred 2011; Parisi et al. 2004; Morton 2008; Wagner and Fernandez-Gimenez 2008). The synergies that these social networks create result in social capital.

Voluntary Environmental Programs (VEPs) It is important to emphasize that while watershed studies identify these three types of watershed collaboratives—agency, mixed, and citizen—as the most predominant arrangements operating in the watershed-policy domain (see Bidwell and Ryan 2006; Koontz and Johnson 2004), scholarship identifies other voluntary partnership arrangements used as policy mechanisms in environmental management. These include environmental voluntary agreements (VAs) such as voluntary environmental programs (VEPs) (Delmas and Terlaak 2001). Although these partnership arrangements are voluntary, the literature reveals important distinctions in VEPs from watershed collaboratives that affect their efficacy in environmental management.

As with watershed agency-based partnerships, VEPs are comprised of institutional actors operating in the public, private, and nongovernmental sphere (Darnell and Sides 2008; Steelman and Rivera 2006). Toddi Steelman and Jorge Rivera (2006, 511) point out that private and nongovernmental actors participating in VEPs predominately represent "industry associations and individual firms, including privately held, state controlled, and publicly traded businesses." Magali Delmas and Ann Terlaak (2001, 44) contend the VEPs fall under the broad rubric of environmental VAs, which they define as "collaborative arrangements between firms and regulators in which firms voluntarily commit to actions that improve the natural environment."

Environmental-policy scholarship suggests that VEPs are designed to address environmental issues identified in specific market sectors such as manufacturing, technology, energy, agriculture, industrial goods, and resorts (Koehler 2007; Steelman and Rivera 2006). The intent of these arrangements is to promote cooperation among regulatory agencies and firms to serve mutual interests (Carmin, Darnell, and Mil-Homens 2003; Steelman and Rivera 2006). VEPs are formal agreements consisting of programs, initiatives, and codes that seek to encourage organizations voluntarily to reduce their environmental impacts beyond regulatory compliance (Carmin, Darnell, and Mil-Homens 2003).

The literature identifies three categories of VEPs: (1) government-sponsored, (2) industry-sponsored, and (3) third-party sponsored (Carmin, Darnall, and Mil-Homens 2003; Darnall and Carmin 2005; Darnell and Sides 2008; Lyon and Maxwell 2007; Steelman and Rivera 2006). By and large, the EPA is the predominant federal regulatory agency that sponsors VEPs (Darnall et al. 2003). Government-sponsored VEPs in the United States are designed to reduce regulatory oversight costs, mitigate administrative burden, increase compliance, and promote innovative processes and procedures within different industries to improve environmental goals (Delmas and Terlaak 2001; Steelman and Rivera 2006). Furthermore, in government-sponored.VEPs, "regulators establish the frame and the basic requirements for participation" (Delmas and Terlaak 2001, 47). Firms participating in these partnerships create their own program initiatives and establish agreements that set "pollution reduction targets" that go beyond regulatory compliance (Delmas and Terlaak 2001, 47). For their participation, regulatory agencies such as the EPA and the US Forest Service provide VEPs funding for research and development as well as technical assistance (Delmas and Terlaak 2001; Steelman and Rivera 2006). Moreover, the literature suggests that VEPs allow firms to gain public recognition as "environmentally friendly" and "green" through organizational recognition programs sponsored by regulatory agencies (Lyon and Maxwell 2007; Darnall and Carmin 2005). Some examples of government-sponsored VEPs are Energy Star, Wa$te Wi$e, and the 33/50 Program (Darnall et al. 2003; Lyon and Maxwell 2007).

Industry-sponsored VEPs are "financed, implemented, monitored and publicized" by industry trade associations (Darnall et al. 2003). Some examples of industry-sponsored VEPs are the Sustainable Forestry Initiative adopted by the American Forest and Paper Association and the Sustainable Slopes Program adopted by the National Ski Area Association (Steelman and Rivera 2006). Finally, third-party-sponsored VEPs are comprised of nonindustry nongovernmental organziations (NGOs) including environmental organizations, industry standardization bodies, and advisory groups (Carmin, Darnall, and Mil-Homens 2003. The International Organization of Standardization (ISO) is one of the most internationally renowned VEPs (Carmin, Darnall, and Mil-Homens 2003).

While there are clear distinctions among VEPs such as composition of stakeholders and sponsorship (see Carmin, Darnall, and Mil-Homens 2003), they all require participating firms to establish environmental performance targets and goals (Darnall and Carmin 2005; Darnall et al. 2003). In addition, partners are required to sign written agreements (i.e., memoranda of understanding) committing to achieving programmatic goals (Darnall and Carmin 2005). An overarching difference between VEP arrangements is in their environmental performance monitoring and reporting protocols (Darnell and Sides 2008; Steelman and Rivera 2006). For instance, some VEPs are self-monitoring, while others use third-party auditors to certify their environmental performance (Darnell and Sides 2008; Koehler 2007).

Despite their appeal for promoting environmental ethics through democratic governance, environmental-policy scholars have questioned the efficacy of VEPs in natural-resource management (Darnell and Sides 2008; Koehler 2007; Lyon and Maxwell 2007). Moreover, some empirical studies conducted by environmental scholars found that overall VEPs had little effect on environmental policy goals (Darnell and Sides 2008; Lyon and Maxwell 2007). Steelman and Rivera (2006, 508) contended that "[c]ommitment to policy goals is necessary to provide direction for envirомental governance." Furthermore, they suggested that the efficacy of VEPs rest on whose interests they serve. Collaborative partnerships are based on common interests shared among partners. In that vein, the effectiveness of environmental collaboratives hinged on the alignment of community interests which include a diverse set of constituents such as business firms, government, local communities, and environmentalists (Steelman and Rivera 2006).

The literature on VEPs reveals four significant differences between VEPs and grassroots watershed partnerships. First, and probably the most striking, is the issue of sponsorship. JoAnn Carmin, Nicole Darnall, and Joao Mil-Homens (2003) characterize sponsorship as the agent(s) that develops, administers, and finances the program. Across all three VEPs, sponsorship centers on institutional actors: governmental, industry, or NGOs. Conversely, grassroots watershed partnerships are developed, administrated, and financially supported by citizens for citizens to address local environmental watershed issues (McKinney and Harmon 2004, Weber 2000). Second, there is the issue of scale and nature of the problem. Environmental issues in grassroots collaboratives are narrow in scope and locally centered, whereas VEPs generally target technically complex environmental issues such as reducing greenhouse gas (Steelman and Rivera 2006).

Third, VEPs focus on achieving superior compliance performance by creating market-sector environmental best-management practices (BMPs) (Darnall and Sides 2008), whereas grassroots watershed partnerships primarily focus on environmental conservation and preservation (Hardy and Koontz 2009). Thus, the intention behind their actions is motivated by a different set of organizational goals. Finally, there is the issue of stakeholder involvement and decision making. In all three VEP arrangements, stakeholder involvement with regard to program ideas consisted of high-level institutional input (Darnall et al. 2003). In contrast, decision making and stakeholder involvement in grassroots watershed organizations seek input from a diverse set of stakeholders that include citizens, local government, environmental groups, and local businesses (McNamara, Leavitt, and Morris 2010).

SOCIAL CAPITAL

In contrast to other forms of capital, James Coleman (1988) argues that social capital is inherent in social structures. It manifests through the interactions and experiences among actors. Moreover, Coleman (1988, S98) underscores that social capital "is not lodged either in the actors themselves or in the physical implements of production."

There are two dominant perspectives found in the social-capital literature. The first perspective suggests that social capital is created through trust and norms of reciprocity (Putnam 2000; Simmel 1950, as cited in Kenney 2000). Through this view of social capital, the behaviors of actors (individual or institutional) are predicated on the notion that the willingness to do *X* is based on the others' willingness to do *Y* (Coleman 1988). In this respect, social capital functions as a transactional mechanism; whereby agreements are based on intangible values such as expectations, trustworthiness, and legitimacy (Coleman 1988; Trachtenberg and Focht 2005).

The second perspective views social capital in the context of networks, whereby the relationships of actors over time form social structures (Lin 2001, as cited in Floress, Prokopy, and Allred 2011). Through interaction these social networks build trust that produces social capital (Floress, Prokopy, and Allred 2011; Leahy and Anderson 2010; Putnam 2000). A large amount of research has been written about the importance of social capital in natural resource management (see Anderson 2011; Leach and Sabatier 2005; Leahy and Anderson 2010; Parisi et al. 2004; Wagner and Fernandez-Gimenez 2008). However, the notion of social capital and its influence on environmental outcomes continues to be an ongoing discourse among scholars (Floress et al. 2011; Leach 2006; Leach and Sabatier 2005; Parisi et al. 2004).

The literature suggests that social capital is an input to the collaboration process, as well as an output (Thomson and Perry 2006; Wood and Gray 1991). On the one hand, research suggests that social capital is the impetus that leads to collective action, thus an input (Parisi et al. 2004; Leach and Sabatier 2005). On the other hand, scholars contend that the dynamics created by the collective action of networks can generate social capital, therefore an output (Leach and Sabatier 2005; Leahy and Anderson 2010; Parisi et al. 2004). From a systems standpoint, collaboration scholars view social capital as a continual feedback loop in the collaboration process (Leach and Sabatier 2005; Wagner and Fernandez-Gimenez 2008). Using these two perspectives, the following discusses the emerging themes found in the literature on watershed collaboration and social capital.

With respect to trust, Zev Trachtenberg and Will Focht (2005) argue that stakeholders' participation is influenced by trust judgements. Furthermore, they contend that trust judgements fall into two dimensions, social trust and official trust. Social trust refers to the level of trust among stakeholders, where trustworthiness is based on the perception that other stakeholders will cooperate in the policy process. Official trust refers to the level of trust between stakeholders and policy officials, whereby the trustworthiness is based on stakeholders' perceptions of how well officials "honor their responsibility to act as stewards of stakeholders' interests" (Trachtenberg and Focht 2005, 91).

Trust judgements move along a continuum, ranging from deference to distrust of others (Trachtenberg and Focht 2005). Trachtenberg and Focht (2005) point out that stakeholders' willingness to collaborate can be influenced by social and official trust. For example, if stakeholders perceive that their interests are not protected by officials or other stakeholders they move toward vigilance. Therefore, they may demand to be included in the policy process. Conversely, if stakeholders share common interests and goals between officials and other stakeholders, this creates deference—thus a willingness to work together.

Trachtenberg and Focht (2005) discuss the concept of legitimacy and its impact on watershed collaboratives. The authors argue that the conditions of procedural and substantive legitimacy must be present to motivate stakeholders to participate in watershed collaboration. Procedural legitimacy refers to the notion that watershed policies are governed by ethical standards and that the deliberation process is fair and just (Trachtenberg and Focht 2005). Moreover, procedural legitimacy holds that the deliberation process balances the interests and concerns of all relevant stakeholders.

On the other hand, substantive legitimacy refers to policy outcomes and whether the policy is justified (Trachtenberg and Focht 2005). In this context, "policy outcome" refers to how the policy improves or diminishes the welfare of stakeholders, which includes social, economic, and environmental factors. Within the framework of substantive legitimacy, Trachtenberg and Focht (2005) provide three criteria: (1) the policy outcome must improve the welfare of individuals other than governmental actors; (2) the policy outcome must include respecting stakeholders' rights; and (3) the policy outcome must be fair in its distribution of stakeholders' welfare. Furthermore, Trachtenberg and Focht (2005) point out that attaining substantive legitimacy in watershed policy is tenuous because of the complexities and unpredictability of watershed issues.

Sarah Cline and Alan Collins (2002) contend that watershed associations that garner social capital are able to shore up support of watershed activities through citizen volunteers. Kristin Floress, Linda Prokopy, and Shorna Allred (2011) and others argue that social capital is essential in community-based collaboratives because it provides access to other resources such as financial, human, political, and technical capital. Floress, Prokopy, and Allred's (2011) study of three local watershed groups in the Midwest used the network perspective to examine the influence of social networks on human capital (i.e., professional skills and scientific expertise). A key finding in the study showed that community-based collaboratives can build human capital through vertical and horizontal "network bridges." Network bridges are created through group members' external connections (Floress, Prokopy, and Allred 2011). They found that external ties can create a "snowball effect," thus creating a chain of human capital. In this regard, Floress, Prokopy, and Allred (2011) contend that watershed groups need to incorporate a "networking minded" spirit in group members.

Along with social capital, scholars contend that "collective efficacy" depends on local socioeconomic and spatial conditions (Irvin and Stansbury 2004; Parisi et al. 2004). Edward Weber's (2000) research on grassroots watershed collaboratives in Montana found that the majority of core members were college educated, held professional jobs, and earned above the national median income. Domenico Parisi's et al. (2004) examination of 324 localities in Mississippi identify level of education as a key driver in "community environmental activeness." Furthermore, Parisi et al. (2004) argue that in low-income areas "community environmental activeness" is difficult because citizens living in these neighborhoods feel disenfranchised from the community at large because they struggle to meet basic human needs (i.e., shelter, clothing, and food).

Edward Weber (2009) found that citizens will engage in collective action to protect the environment when they share a commitment to "place." Weber's (2009, 318) case study of the Blackfoot watershed found that a "common vision of place" allowed stakeholders with competing interests to put aside their differences and focus on what they shared, namely, their love for the Blackfoot area and preserving the environment for future generations. Parisi et al. (2004) contend that "spatial identity" (size and regional location) of the community is a factor in environmental activeness. In their study, Parisi et al. (2004, 109) found that in communities with populations of 2,500 or less there were "significantly lower levels of environmental activeness than larger places, even after controlling for other social and economic conditions." They attribute this to inadequate leadership, lack of access to resources, and the lack of strong social network bridges within the community.

Cheryl Wagner and Maria Fernandez-Gimenez's (2008) case study of eight community-based environmental collaboratives in Colorado identified factors that facilitated social capital. First, they identified impartial deliberations. They found that establishing ground rules that considered all stakeholder's viewpoints and inputs equally fostered cooperation and built trust. Second, they identified the development of open forums between agency staff, residents, and group members as beneficial in the collaboration process. In this regard, the authors contend that venues promoting the exchange of viewpoints, information, and ideas foster trust, respect, and understanding among stakeholders. Their study also showed that certain activities can diminish the level of social capital in groups. They note that "controversial group actions" such as the withdrawal of promised resources or the exiting of key stakeholders affected the level of social capital. In their research, Wagner and Fernandez-Gimenez (2008) identify eight variables that foster and enhance social capital in collaboratives. They found that in community-based natural resource collaboratives the following variables are essential:

- *Respect* and *understanding*—respecting and listening to the viewpoints of all partners facilitates mutual understanding, thus laying the foundation for trust and social capital;
- *Empathy*—mutual understanding among partners leads to collective reverence towards diverse perspectives, which fosters empathy;
- *Commmitment* and *continuity*—commitment to the partnership and consistent participation among members leads to norms of reciprocity;
- *Predictability* and *dependability*—the ability to depend on partners and predict their actions creates trustworthiness and;
- *Transparency*—open communication concerning partner's motives, objectives, and actions instills transparency.

Tomas Koontz and Craig Thomas (2006) maintain that social capital creates as well as supports the infrastructure of collaboration. Some suggest that the effectiveness of watershed partnerships depends on the outputs (i.e., agreements, documents, and projects) produced by the collaboration process (Koontz and Thomas 2006; Leach et al. 2002). The following section discusses the prevailing themes found in the literature on watershed outputs and their implications to outcomes.

WATERSHED OUTPUTS

Koontz and Thomas (2006, 115) argue that watershed outputs are the "intermediary causal mechanisms between collaboration process and collaborative outcomes." Research suggests that performance measures for watershed collaboratives primarily use output measures rather than outcomes (Koontz and Thomas 2006; Mandarano 2008). They attribute this to the costs associated with monitoring outputs as well as the confounding variables that affect environmental conditions (Lubell et al. 2005; Mandarano 2008; Thomas 2008). Some examples of watershed outputs include watershed plans and agreements, scientific reports, and water-quality standards (Lubell et al. 2005; Mandarano 2008).

The literature suggests that formulating watershed plans is essential for grassroots environmental collaboratives because funding sources often require watershed partnerships to present their watershed plan in order to receive grants (Koontz and Thomas 2006). Leslie Duram and Katharin Brown's (1999) examination of 126 federally funded watershed initiatives found that plans vary among partnerships and depend on the environmental issues of the watershed. However, their study identifies several commonly shared characteristics in watershed planning:

- watershed planning incorporates a wide variety of input from a diverse set of stakeholders;
- issue framing and objectives are decided by consensus; and
- plans are influenced by the characteristics of the watershed populations (i.e., rural, residental, agriculture).

Once the plans are formulated, agreements must be forged in order to accomplish the plan's objectives (Leach, Pelkey, and Sabatier 2002). Some agreements simply provide outlines for the partnership's objectives, whereas in more comprehensive agreements stakeholders agree to implement specific action plans (Leach, Pelkey, and Sabatier 2002). Scholars suggest that the level of sophistication in agreements may depend on the the composition of stakeholders and their environmental agendas (i.e., governmental, nongovernmental, and citizens) (Hardy and Koontz 2009; Imperial 2005; Margerum 2008). With regard to evaluating the agreements, Margerum (2002) contends that quality agreements articulate clear goals and expectations.

Scholars contend that there is a positive correlation between forging agreements and trust and social capital (Lubell et al. 2005; Trachtenberg and Focht 2005; Mandarano 2008). Other factors that influence the ability to reach agreements is the age of the partnership and the severity of the watershed issue (Leach and Sabatier 2005). Willam Leach and Paul Sabatier (2005) found that in partnerships of less than three years, the primary driver for reaching agreements was a watershed crisis, whereas in older, more established partnerships, the driver was trust among stakeholders. Leach and Sabatier (2005) maintain that repeated interaction among stakeholders builds trust, thus facilitating reaching agreements.

Koontz and Thomas (2006) point out that forging agreements does not necessarily equate to projects that improve watershed conditions. Therefore, they assert that measuring a partnership's activities provides an indication of the effectiveness of the collaboration process. These activities include restoration projects and education and outreach programs. Mark Imperial's (2005, 285) examination of six watersheds found that partnerships' activities directly relate to their "ecological settings, environmental problems, institutional environments and situational histories."

Scholars assert that linking outputs to watershed improvement is particularily tenuous in grassroots partnerships (Koontz and Thomas 2006; Lubell 2004a). Although information on the general condition of the watershed is available to local watershed partnerships, William Leach, Neil Pelkey, and Paul Sabatier (2002) found that few partnerships have the resources to develop and/or support monitoring the impact of outputs on long-term ecological changes in the watershed. It is thus difficult for researchers, policy makers, and public administrators to make inferences about the relationship between watershed partnerhips' outputs and actual environmental improvements (Lubell et al. 2005; Kenney 2000; Margerum 2002). Consequently, researchers use surrogate measures to evaluate watershed outputs that are easily qualifiable. These measures can include the number of restoration projects completed, the number of changes to environmental practices, and the number of outreach programs conducted (Campbell, Koontz, and Bonnell 2011; Lubell et al. 2002, 2005; Margerum 2007). While these are tangible measures, they fail to provide evidence of improved environmental conditions.

Watershed Outcomes

As watershed partnerships gain prominence in managing natural resources, contemporary scholarship is focusing on examining the efficacy of these collaboratives (Campbell, Koontz, and Bonnell 2011; Conley and Moote 2003; McKinney and Field 2008; Leach and Sabatier 2005; Lubell 2004b). An emerging question that appears in the literature is, what criteria should be used to evaluate the efficacy of watershed partnerships? Scholarship identifies two broad types of watershed outcomes, environmental and social (Koontz and Thomas 2006; Mandarano 2008). Environmental outcomes entail changes to ecological systems such as improvement in water quality and habitats, while social outcomes refer to changes in social structures such as increases in civic engagement, volunteerism, and trust (Morton 2008; Thomas 2008; Leach and Sabatier 2005).

The literature identifies two key factors that indicate effectiveness in environmental outcomes. The first is the implementation of restoration projects that target activities such as abating nonpoint source pollutions, increasing vegetation, or changes to in-stream flows (Leach, Pelkey, and Sabatier 2002). The second are changes in land use through acquisition, zoning, and easements (Leach, Pelkey, and Sabatier 2002; Mandarano 2008). In addition, Lynn Mandarano (2008) contends that improvements in ecosystem biodiversity provides evidence of successful watershed projects. However, Mandarano (2008) notes that detecting changes in environmental parameters requires ongoing monitoring, which is dependent on resources available. Others identify changes in the level of environmental litigation as an indicator (Conley and Moote 2003). Craig Thomas (2008) and others caution that researchers need to use an evidence-based approach to evaluate environmental outcomes in watershed partnerships. In this regard, Thomas (2008) argues the evidence of outcomes should drive the selection of outputs, rather than using outputs as surrogates for outcomes.

We take some issue with the characterization of project implementation as a watershed outcome measure. An important distinction between outputs and outcomes is the temporal "distance" from the activity under study—in this case, collaboration. While program implementation is an ongoing process, it also begins at the point at which resources are committed to a decision. In short, implementation begins in the near term, and is thus better termed an output of collaboration. On the other hand, outcomes take significantly longer to manifest, and may not be obvious (or detectable) for a much longer period of time. A project to remove contaiminated sediments from a riverbed begins when workers begin their activities, but changes to water quality may not be detectable for months or years after the initial work has begun. Likewise, the implementation of an education program for high-school students may not produce measurable changes in behaviors for several years following the treatment. Therefore, we treat the implementation of projects as an output of collaboration, rather than a long-term outcome.

With respect to social outcomes, the literature identifies several indicators of watershed collaboration effectiveness, including changes in environmental values and ethics, changes in behaviors, social capital, trust, intellectual capital, and perceptions of environmental conditions (Campbell, Koontz, and Bonnell 2011; Leach and Sabatier 2005; McNamara, Leavitt, and Morris 2010; McKinney and Harmon 2004; Weber 2009). Regarding environmental values and ethics, scholars found that public education and outreach projects offering training and workshops in watershed protection and conversation are benefical in creating public awareness of local watershed issues, which ultimately leads to better stewardship of the environment (Leach, Pelkey, and Sabatier 2002; McNamara, Leavitt, and Morris 2010; Weber 2000).

Joseph Campbell, Tomas Koontz, and Joseph Bonnell (2011) attribute the adoption of BMPs as an indicator of effective social outcome. These practices include "integrated pest management, cover crops, [and] vegetated buffer strips" (Campbell, Koontz, and Bonnell 2011, 1129). Their analysis shows that farmers active in watershed partnerships are more likely to adopt BMPs than nonactive farmers. In this vein, Lois Morton (2008) suggests that informal social structures, such as local community collaboratives, facilitate the exchange of local and technical knowledge among stakeholders. Morton (2008) found that in rural communities, informal networks are more effective at disseminating new techniques and skills; thus old practices are replaced by BMPs.

In regard to stakeholders' perceptions of watershed conditions, the literature suggests that participants' perceptions, while important, may be biased (Koontz and Thomas 2006). Koontz and Thomas (2006) and others find that positive perceptions of watershed conditions among collaborative groups are in part due to the solidarity that is generated through collective action. Therefore, scholars attribute collective efficacy in watershed collabortives to the "halo effect" rather than substantive evidence of improved environmental conditions (Koontz and Thomas 2006; Lubell et al. 2005; Trachtenberg and Focht 2005). Koontz and Thomas (2006) assert that while social outcomes such as increased civic engagement (i.e., social capital) are important, the true measure of success is their impact on environmental conditions. In addition, the literature reveals that assessing the efficacy in watersheds requires substantial resources as well as long time horizons (Sabatier, Weible, and Ficker 2005; Kenney et al. 2000).

Putting It All Together

The literature developed in the preceding discussion lays the foundation for our model of grassroots environmental collaboration. The contextual elements, both national and local, form the setting for the collaborative activity. The collaboration process takes place within this context, and consists of actors, resources, rules, and governance structures. The two proximate results of the collaboration process are outputs, which are agreements or projects, and social capital. Successful outputs lead to watershed outcomes, or measurable changes in the perceived shortcomings within the watershed. Outputs also serve to increase social capital, as citizens feel a sense of accomplishment at having achieved agreements and/or started projects. Figure 2.2 presents our complete model, including the components of each major conceptual element.

Both social capital and watershed outcomes serve to alter the original context in which the collaboration activity occurs. As social capital in a

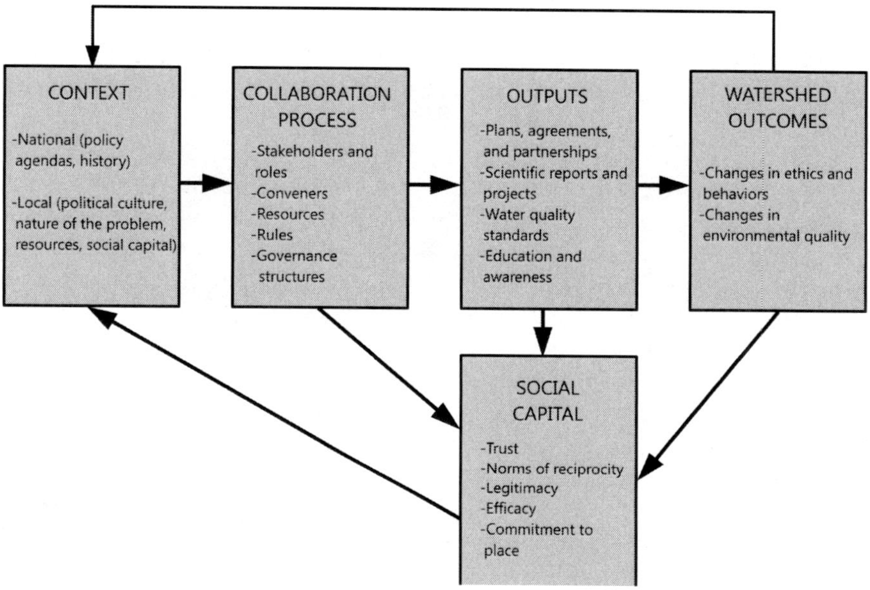

Figure 2.2. An Enhanced Model of Collaboration

community builds, the level of support for collaboration increases, and also serves to bring new citizens into the process. Likewise, demonstrable positive (or negative) changes in the watershed alter the demands for changes, and can also serve to strengthen both visibility and support for the efforts to address water quality. Thus a cyclical process is formed, just as envisioned by Easton (1965) in his original systems model of the policy process.

STUDY METHODS

To better inform the academically minded readers of this work, this section of the chapter discusses the methods employed for this study. In order to make this discussion more understandable to a nonacademic audience, we will provide some additional explanation of our choices.

The methodological approach employed for this research is a qualitative case study. Robert Yin (1994) describes several approaches to case-study research; the method we adopt for the study is an embedded, multiple-case design. Our work seeks to be explanatory in nature. Explanatory approaches, according to Yin (1994, 9), occur when "a 'how' or 'why' question is being asked about a contemporary set of events over which the investigator has little or no control." In a controlled experiemental environment, the investigator exercises control over a range of variables, and can thus focus attention on a specific factor (or variable). Because we are studying a complex set of human interactions as found in natural settings, experimental methods are not appropriate for this research.

Case studies are also useful when there are more variables of interest than there are data points to measure. Case studies allow researchers to investigate complex interactions between many factors within a setting, and to delve into these complex interactions to discern not only patterns of behavior, but to better understand the observed patterns. To accomplish this, we rely on multiple sources of evidence. Finally, we benefit from the development of a theoretical model, and the propositions we derive from that model, to inform our efforts in both data collection and analysis (Yin 1994, 13).

Because this is an embedded multiple-case design, we investigate more than one unit of analysis. At the first level, our unit of analysis is the grassroots environmental organization and the collaborative partners they develop to accomplish their missions. However, we also place a great deal of emphasis on the individual actors within each grassroots structure, with particular attention to their motivations, contributions, expertise, and actions within the organization. We also examine both the networked structures each organization develops to enhance its capabilities, as well as the patterns of behavior between the network actors.

Data Sources

Several sources of data were utilized for this study. The following sections describe those sources, and the collection methods employed.

Interviews

Our major source of data was in-person interviews of citizens involved in each of our three case organizations, as well as representatives of local governments, state and federal agencies, and other nonprofit environmental groups, conducted between January and October 2012. We utilized a snowball sampling method. First, we identified the executive directors of each of the three organizations, and met with each separately to secure permission to conduct the case study. We asked each of the three directors for a list of people important to the organization, contacted each of the people on the initial lists, and arranged interview times with each. We also asked for contacts in local governments and state and federal agencies, and relied on several of our personal contacts in these organizations as well.

We used semistructured interview protocols developed from our theoretical framework. Each interview was conducted in a place convenient to the interviewee, and ranged from offices to restaurants, university conference rooms, and private homes. The research team made notes about the setting of each interview. After a brief discussion of the purpose of the research, and the receipt of a signed consent form from the interviewee,[7] we began the interview. All interviews were recorded (with the permission of the interviewee), and the interviews were subsequently transcribed. Each interview lasted between sixty minutes and three hours, depending on the availability and interest of the interviewee. The use of a semistructured interview protocol allowed the interviewers to follow up on new (or unexpected) issues raised by the interviewee. Depending on the role (and willingness) of the interviewee, some people were interviewed multiple times. At the conclusion of each interview, the interviewee was asked to name other individuals they considered to be important to the efforts of the group. These people were subsequently contacted and interviewed. Names were often mentioned by multiple interviewees, but after several rounds of interviews, the names we were given by new interviewees matched names already on our lists.

We also conducted interviews with several individuals not members of these organizations, but with whom these groups regularly interact. These individuals included representatives of industry; local, state, and national government; regional authorities; and other environmental organizations. In many cases, these people collaborate with one or more of our case organizations, but several were interviewed as a means to gain a sense of the broader context in which the collaboration takes place. In all, we conducted about eighty-five interviews. A list of the people interviewed, and their organizational connections, appears in the appendix.

The transcripts of the interviews were reviewed by the research team, along with our written notes from the interviews. This review focused on the identification of important themes, events, and elements of each case. Recordings of the interviews were also reviewed to discern nuances of tone, mood, and context for comments deemed to be insightful or important.

Document Analysis

Because all three of our case organizations are 501(c)3 (nonprofit) entities, each produces (and retains) a series of important documents. With the permission of each organization, we were able to review meeting agendas, meeting minutes, reports, planning documents, watershed action plans, brochures, memos, and promotional materials generated by each group. In addition, we conducted a thorough analysis of each group's website, and performed a Google search on each organization to identify other relevant documents for each. The documents were reviewed for the presence of themes identified from the interviews, as well as themes not raised in the interviews. We also used these documents as sources of factual information for our case analyses.

A related effort was conducted by examining the archives of the *Virginian-Pilot,* the major newspaper in the region. We began our search for articles published in 1992, the year of the formation of the oldest of our three cases (the Elizabeth River Project), up to September 2012. This search returned some eight hundred articles in total, of which about a third were announcements of upcoming events. The articles were analyzed for factual content, as well as for themes consistent with those previously identified.

Direct Participant and Nonparticpant Observation

Study team members attended public functions held by each of the three case organizations. These events ranged from public talks to fests and river cleanup events. The events took place between April and September 2012. The research team members observed the interactions between members of the organizations, and recorded notes about the kinds of activities that constituted each event. In addition, research team members were able to attend staff and committee meetings of each organization as nonparticipant observers (see Johnson and Joslyn 1991). Notes were made about the nature of the interactions and the conduct of the meetings, but not about the substantive content of the meetings.

We also had several opportunities to engage directly in activities for each of the case-study organizations. These events included a trip to collect water samples on the Nansemond River, an inspection tour of the Paradise Creek Park in Portsmouth on the Elizabeth River, a tour of several wetlands restoration projects on the Elizabeth and Lafayette Rivers, and a hands-on experience on an oyster boat on the Lynnhaven River. These events gave the research team the opportunity to experience firsthand the activities of the groups, and to gain a fine-grained appreciation for the kinds of work in which these groups engage.

Analysis Methods

We employed content analysis techniques (see Nachmias and Nachmias 1996) to derive meaning from our data. This initial analysis provided a means to determine whether our initial (and impressionistic) thematic analysis was borne out by a more systematic method. We then matched the themes identified to the variables identified in our conceptual framework, and examined the hypotheses and propositional statements suggested by the framework. We then applied the separate document analyses conducted to the same framework to determine whether our initial data could be triangulated. Finally, we applied our observational data as further triangulation; these procedures also addressed questions of construct, internal, and external validity (see Yin 1994).[8] While our study is not specifically ethnographic in nature (see Altheide and Johnson 1998; Yin 1994), we also reviewed our data as a means to understand the system of meanings present in each social setting (see Geertz 1973). The research team met regularly to discuss findings, review data, and ensure that we maintained a high degree of intercoder reliability.

Layout of the Case Studies

The next four chapters of this book outline the broader setting of our study, and detailed discussion of each of our three cases. The three case studies are organized around the elements of the framework developed in this chapter, as a means to facilitate comparison across the three cases. Chapter 7 draws some broad comparisons across these categories, and draws some conclusions about the efficacy of the framework.

NOTES

1. Easton (1965) conceived of the policy process as a process consisting of three major parts, connected by a feedback loop. For Easton, demands and support constituted the inputs to the process. The inputs fed into a "black box" policy process, which in turn created outputs (new policies). All of this activity took place in a larger environment, which both provided the inputs and received the outputs. An output from this process would necessarily modify the environment in some way, thus altering the inputs available for subsequent policy action.
2. The nature of the outputs produced necessarily drives the kinds of outcomes possible in the watershed.
3. Significant parts of this section were written by Luisa Diaz-Kope, a doctoral student at Old Dominion University, as part of a Directed Research course during Summer 2012. We are grateful for her contribution to this chapter.
4. The Water Quality Act also established the Federal Water Pollution Control Administration as the agency responsible for water quality. Up to this point, responsibility for water pollution had been a function of the US Public Health Service (Caulfield 1989).
5. Elazar (1984, 115) notes explicitly that these three labels are meant to be descriptive, not evaluative.
6. The levels include operational, collective, and constitutional; see Hardy and Koontz (2009) for a more detailed explanation of these levels.
7. This research was approved by the Institutional Review Boards at both Old Dominion University and Virginia Wesleyan College, and the research was conducted in accordance with the fundamentals of Responsible Conduct of Research (RCR).
8. Because we conducted three separate case studies, we cannot directly address issues of reliability. However, the research team took great care to document our procedures and techniques during the study.

THREE

The Chesapeake Bay

Local Efforts, Regional Connections

> *As efforts to safeguard the Bay's future exact a greater cost in terms of taxpayer dollars and the diminution of local authority and traditional Bay cultures, the challenge will be to fashion a public consensus that everyone has a stake in the Bay.*
>
> —Davison et al. (1997, 229)

Despite the Bay's productivity, natural splendor, and economic importance to the region, and despite almost three decades of regional restoration efforts, almost all of the estuary and its tidal waters remain listed as impaired under the Clean Water Act. While the Bay as a whole often receives enormous press and attention, whether it is ultimately restored to health depends in large part upon the success of both regulators and grassroots environmental groups to achieve local water-quality gains—stream by stream, river by river. Thus the lessons learned from local collaborations have broader implications for environmental protection than it first may seem. Our study reveals that environmental groups in Hampton Roads—and groups like them around the watershed—are crucial pieces of the larger Bay-restoration puzzle, and are likely to play an ever-increasing role.

In addition to having an impact on Bay restoration generally, the local collaborations we studied are also affected by being located at the mouth of the Bay and within the Bay watershed. No estuary in the world has been studied and analyzed as the Bay has been, and the quality and amount of science that has resulted from this work is staggering in depth and scope. Meanwhile, organizations such as the National Oceanographic and Atmospheric Administration (NOAA) and the US Army Corps of Engineers (USACE) have a large presence in Hampton Roads because of the natural harbor created as the James River joins the Nansemond and Elizabeth Rivers and drains into the Bay at the Bay's mouth. As a result, scientific expertise and financial resources dedicated to restoring the Bay often support—directly and indirectly—local grassroots efforts. Bay policies and politics, which are likewise impressive in their depth and scope, also inevitably inform each group's direction and mission.

This chapter first provides an overview of Hampton Roads, the area that is home to each of the watersheds we studied, and then discusses the Chesapeake Bay and its importance to the larger region in which Hampton Roads sits. It then describes the Bay-wide regulatory, scientific advisory, advocacy, and funding contexts, and includes examples of the relationships or "linkages" that groups in Hampton Roads have with each other and with state, regional, and national groups. The chapter then turns to regional collaboration in the Bay. In this section, we describe how a regional approach to restoration evolved, and then compare the regional lack of progress with the very concrete progress that was made in the Lynnhaven River to improve its water quality by decreasing bacteria levels. The comparison leads us to conclude that the most likely road to improved water quality involves both regulatory and voluntary approaches, as well as increased appreciation for the important role local watershed collaborations play in Bay restoration efforts.

THE SETTING: HAMPTON ROADS

Hampton Roads, one of the world's largest natural harbors, sits at the mouth of the Chesapeake Bay, the largest estuary in North America. The James, Nansemond, and Elizabeth Rivers join to create the harbor channel, along with several, smaller rivers. The waterway is so important to southeastern Virginia that the entire region now bears its name. Hampton Roads—the region—encompasses the cities of Norfolk, Virginia Beach, Newport News, Chesapeake, Hampton, Portsmouth, Suffolk, and Williamsburg, among others. Its population of about 1.7 million makes Hampton Roads the thirty-sixth-largest metropolitan statistical area in the country (US Census Bureau 2010).

Given the confluence of the harbor, the Chesapeake Bay, and the Atlantic Ocean, the Hampton Roads region's economic base has long been port-related, a natural site of maritime and military activity. The world's largest naval station, Naval Station Norfolk, and the world's largest shipyard, Northrop Grumman Newport News, call Hampton Roads home. The Norfolk Naval Shipyard in Portsmouth is the US Navy's largest and oldest industrial facility. The US Coast Guard, Air Force, Marines, and Army also have facilities in Hampton Roads, giving the area the distinction of having the largest concentration of military bases in the world (Norfolk Department of Development 2011).

Private commerce also thrives thanks to the region's waterway resources. There are major container and bulk cargo terminals located in the area. As one of the largest cargo ports in the United States, railroad giants Norfolk Southern and CSX and numerous trucking companies have a large presence in the area, and Hampton Roads is the nation's leading exporter of coal (Virginia Port Authority 2012). A large number of manufacturing, processing, and storage companies also support the shipyard industries. Tourism likewise thrives in Hampton Roads. An estimated 2.75 million tourists visit Virginia Beach yearly (Virginia Beach Economic Development 2012). A recently completed cruise ship terminal in Norfolk brings forty thousand to sixty thousand passengers to the area annually (McCabe 2010). Sport fishing, sailing, and water recreation are popular.

Although many of Hampton Roads' rural areas have been lost and continue to be lost to suburban development, agriculture and commercial fishing also contribute to its economy and culture. Farming is a significant element of the economy in the city of Virginia Beach, and peanuts were once a major industry for the city of Suffolk, the home of Planters Peanuts. The world's largest pork producer and processor, Smithfield Foods, Inc., is headquartered in Smithfield, Virginia, and Newport News boasts a thriving seafood industry. In total value of seafood landed, Hampton Roads is the sixth-richest port in the United States (Newport News Department of Development 2011).

With all of this activity, it should be no surprise that the region's waterways are polluted and have been for a very long time. To compound the problem, Hampton Roads sits in the Lower James River basin, on the receiving end of the James River, which has the highest level of nutrient pollution in the state due to outdated sewage-treatment plants, suburban sprawl, and agricultural runoff (Chesapeake Bay Foundation 2005). Much of this pollution finds its way into Hampton Roads, a network of waterways already burdened by the pollution produced by its own industries, development, and people. The primary impairments that plague the area include bacteria, mercury, polychlorinated biphenyls (PCBs), and nutrient pollution (excessive levels of nitrogen and phosphorus) (Chesapeake Bay Foundation 2005). The sources of this pollution include sewage-treatment plants, industrial plants, stormwater runoff, and agricultural runoff. Development also has resulted in extensive wetlands loss.

The confluence of waterways at the mouth of the Chesapeake provides a unique opportunity to compare grassroots river restoration efforts that are geographically similarly situated and arise from generally similar political and social cultures. This book thus examines in depth three river restoration efforts in Hampton Roads: Lynnhaven River NOW (LRN), the Elizabeth River Project (ERP), and the Nansemond River Preservation Alliance (NRPA). A brief overview of the rivers (see figure 3.1) that are the focus of these efforts follows.

The Lynnhaven River

Located entirely in the city of Virginia Beach, the Lynnhaven River is a tidal estuary that flows into the Chesapeake Bay. Once a productive oyster-harvesting area, the river's "Lynnhaven Oysters" were world famous until excessive bacteria levels forced the Virginia Health Department to close almost all of its shellfish areas. While fecal coliform was the culprit, Virginia Beach's population explosion was the cause. Approximately 225,000 people reside within the Lynnhaven's sixty-four-square mile watershed (Lynnhaven River NOW 2010). Sanitary sewer overflows, failing septic systems, stormwater discharges, and sanitary discharges from boats polluted the river to such an extent that its once treasured resource became inedible.

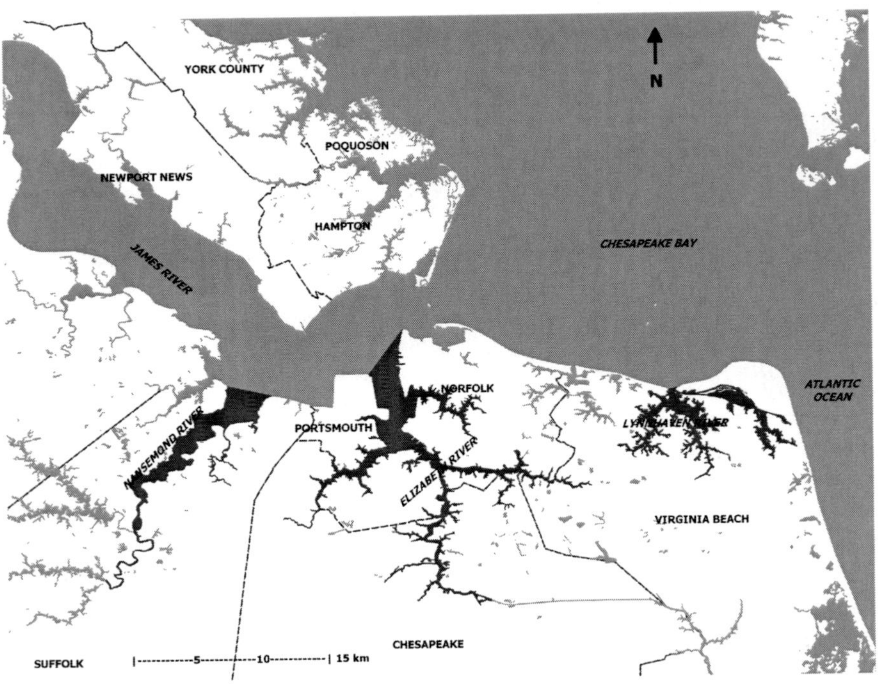

Figure 3.1. The Hampton Roads Region

The Elizabeth River

The Elizabeth River, a six-mile-long tidal estuary, flows into Hampton Roads harbor and into the Chesapeake Bay. While its main branch runs through Norfolk, three major branches to the west, south, and east reach into Portsmouth, Chesapeake, and Virginia Beach, respectively. Significant military and commercial port facilities in Norfolk, Portsmouth, and Chesapeake depend upon the river. Described as the "birthplace of maritime industry in America," the Elizabeth River also holds the distinction of being not only one of the most polluted rivers to flow into the Chesapeake Bay, but also one of the most polluted rivers in the nation (Kobell 2011). Four centuries of industrial and military use have resulted in a highly contaminated river bottom. Some of the world's highest levels of polycyclic aromatic hydrocarbons (PAHs), which cause cancer in fish and pose a risk to human health, are found in nineteen acres of the river's sediment. Indeed, nine hazardous "cleanup" sites line the river's shorelines, with the US Environmental Protection Agency (EPA) designating an entire "remediation team" to oversee toxic remediation efforts on the river. Nutrient pollution from stormwater runoff also plagues the Elizabeth. The fertilizers, herbicides, pesticides, and pet waste found in stormwater runoff generates excessive algae growth, which in turn depletes dissolved oxygen in the water necessary to support marine life (Elizabeth River Project 2008a).

The Nansemond River

Originating in downtown Suffolk, the Nansemond River runs for twenty miles to eventually join the James River and enter the Chesapeake Bay. Its watershed has historically supported agriculture, namely, tobacco, peanuts, soybeans, and cotton. Oyster, crab, and shad fisheries once thrived on the Nansemond as well, but pollution, overfishing, and loss of riparian habitat to development have all but destroyed these resources. Fecal coliform bacteria by means of stormwater runoff is the primary pollutant of the river, resulting from animal waste, aging sanitary sewer systems, and septic systems. Nutrient pollution from fertilizers on agricultural and suburban lands also impairs the river. Wetlands loss compounds the problem, as crucial natural filters that could control runoff are lost to development (Nansemond River Preservation Alliance 2011).

THE CHESAPEAKE BAY: A BRIEF OVERVIEW

The Chesapeake Bay, the second-largest estuary in the world, is renowned for its crabs and oysters, its beauty, and its importance to American culture and history. Its economic importance to the region is significant. The seafood industry alone in Maryland and Virginia contributed "$3.39 billion in sales, $890 million in income, and almost 34,000 jobs to the local economy," while saltwater recreational fishing was estimated to provide "$1.34 billion in sales" and "roughly 11,000 jobs" (Chesapeake Bay Foundation 2009). Two of the Eastern Seaboard's major ports—Baltimore and Hampton Roads—are located on the Chesapeake.

The Bay also has several striking geographic characteristics that make it both fascinating to explore and frustrating to protect. The Chesapeake sits within a six-state, sixty-four-thousand-square-mile watershed, making its "land-to-water ratio (14:1) the largest of any coastal water body in the world" (Davison et al. 1997). This geographic area is home to more than 17 million people. Five major rivers—the Susquehanna, Potomac, Rappahannock, York, and James Rivers—and more than one hundred thousand streams, creeks, and small tributaries drain into the Bay (Chesapeake Bay Program 2012a). The Bay also boasts a very long shoreline—approximately eight thousand miles—in relation to the overall size of its water body. What this means is that the Bay is extremely "branchy," with numerous creeks, bays, inlets, and "guts." Sailors love exploring the Bay as a result, but the Bay's "branchiness" also increases the ability for people to live on the water as well as impact the Bay's water quality by how they use their land (Davison 1997). The Bay is also unexpectedly shallow, with an average depth of twenty-one feet. As the Chesapeake Bay Program (CBP) explains, "a person who is 6 feet tall could wade through more than 700,000 acres of the Bay and never get his or her hat wet" (Chesapeake Bay Program 2012b). The Bay's shallowness is both its source of great productivity—it supports 3,600 species of plant and animal life—and its Achilles heel (Horton and Eichbaum 2003). The Bay does not "flush" pollutants away or dilute them well, allowing local water-quality conditions to become and remain acute. Moreover, while shallow waters allow for sunlight to penetrate the water column, creating a welcoming habitat for underwater grasses, crabs, and shellfish, they also make the Bay naturally susceptible to excess algae growth and the subsequent development of "hypoxic conditions," where dissolved oxygen becomes so reduced in concentration that aquatic organisms cannot survive (Boesch, Brinsfield, and Magnien 2001).

Excess nutrients—nitrogen and phosphorus—introduced into the aquatic ecosystem are the primary causes of hypoxic "dead zones." From 2009 to 2011, only 34 percent of "the combined volume of open-water, deep-water and deep-channel water of the Bay and its tidal tributaries met dissolved oxygen standards during summer months" (Chesapeake Bay Program 2012c). According to the CBP, in 2009, approximately 245.8 million pounds of nitrogen and 16.46 million pounds of phosphorus drained into the Bay from rivers and streams across the watershed. Runoff from agriculture is the largest culprit, contributing 45 percent of nitrogen loads and 44 percent of phosphorus. Point source pollution and urban runoff are also important contributors; their combined contributions almost total the amount contributed by agriculture. Polluted runoff from developed lands is the fastest-growing source of pollution (Chesapeake Bay Program 2012c).

WATERSHED RESTORATION IN HAMPTON ROADS: LINKAGES

Given their location, the watershed collaboratives we study sit within a sophisticated and complicated regulatory, scientific advisory, advocacy, and funding context. In this section, we first provide an overview of the key government actors from federal, regional, and state levels that have the greatest impact on environmental restoration in Hampton Roads. We then turn to key regional and state environmental groups. Throughout our discussion, we include examples of the relationships or "linkages" that groups in Hampton Roads have with each other and with state, regional, and national groups.

Key Government Actors

ERP, LRN, and NRPA each receive support from a wide variety of government actors by means of grants, technical assistance, and information about their waterways and watershed. Some of these same government actors also impose new regulations and policies, take enforcement actions, and supervise complicated cleanups. When it comes to providing funding, expertise, and institutional support for watershed restoration, USACE (or "Corps"), NOAA, and the EPA are the most important federal actors in Hampton Roads. The Virginia Department of Environmental Quality (DEQ) and the Virginia Department of Conservation and Recreation (DCR) are the most important state actors. The CBP is the most important regional group of government actors watershed-wide, while the Hampton Roads Planning District Commission (HRPDC) holds that status in Hampton Roads. We briefly review both the type of support and the regulatory pressure that affect the restoration efforts of the groups we studied. Textbox 3.1 lists the organizations covered in this discussion.

Textbox 3.1

Summary List of Organizations

National
US Army Corps of Engineers
National Oceanographic and Atmospheric Administration
US Environmental Protection Agency

State
Virginia Department of Environmental Quality

Virginia Department of Conservation and Recreation
Old Dominion University
Virginia Institute of Marine Science

Local/Regional
Hampton Roads Planning District Commission
City of Virginia Beach
City of Norfolk
City of Portsmouth
City of Chesapeake
City of Suffolk

Private/Nonprofit/Partnerships
Chesapeake Bay Program
Virginia Conservation Network
Chesapeake Bay Foundation
Choose Clean Water Coalition
Virginia Environmental Endowment
Elizabeth River Project
Lynnhaven River NOW
Nansemond River Preservation Alliance

The US Army Corps of Engineers

As "the nation's environmental engineer," the Corps' presence in Hampton Roads reflects its multifaceted mission that includes, among other things, conducting and supporting ecosystem restoration, regulating waterways and wetlands, and cleaning up contaminated military sites. With a district office of more than four hundred civilians and Army officers located in Norfolk, the Corps has been, and continues to be, a steady source of volunteer expertise for the grassroots collaborations we studied (Hume 2012). For example, Joe Thomas, ERP's second president, was retired from Corps, after serving as a colonel in charge of Norfolk's district office (Thomas 2012). In its official capacity, the Corps has developed restoration plans for both the Lynnhaven and the Elizabeth Rivers, with the HRPDC playing a coordinating role (Tribo and Katchmark 2012). Indeed, Clay Bernick, the City of Virginia Beach's director of environmental services, credits work initiated by the Corps in the early 2000s for increasing environmental restoration coordination among Norfolk, Virginia Beach, Chesapeake, and Portsmouth by means of the HRPDC (Bernick 2012).

The Norfolk District's "native oyster restoration efforts" have focused on both the Lynnhaven and Elizabeth Rivers as well, although its work in the Lynnhaven has been more extensive to date. In the Lynnhaven, the Corps—partnering with the City of Virginia Beach, LRN, the Virginia Institute of Marine Science (VIMS), the Chesapeake Bay Foundation

(CBF), and NOAA—has constructed twelve sanctuary oyster reefs over fifty-four acres of river bottom and seeded them with oyster spat, at a cost of $4.7 million (Lynnhaven River NOW 2012a). In 2010, David Schulte, a marine biologist with the Corps, accepted on behalf of the Norfolk District the Coastal America Partnership Award, which included recognition from the White House, for this work (Harper 2010).

The Corps' relationship with ERP has been particularly important, although it has changed over time. In 1998, in what was described as a "milestone" for ERP, the Corps received $2.4 million from local, state, and federal officials to conduct a three-year study on rebuilding shoreline wetlands and ridding river sediment of toxic chemicals (Harper 1998). The funding and support for the study represented a joint effort among the mayors of Norfolk, Virginia Beach, Portsmouth, and Chesapeake, as well as state and federal officials. While the initial thought was that the Corps would conduct environmental restoration in the Elizabeth as an expansion of its mission, a different approach was ultimately taken (Jackson 2012b). The pivotal event occurred in 2001 when Maersk Sealand, the world's largest shipping line, proposed building a large commercial port on the Elizabeth River (O'Brien 2001). EPA and the US Fish and Wildlife Service originally opposed Maersk's permit before the Corps. After negotiation among the federal agencies involved, the Virginia Department of Environmental Quality, ERP, and Maersk, Maersk agreed to increase its environmental compensation funding from $900,000 to $6.5 million, with $5 million to be directed to a trust fund established to support ERP-recommended projects (Harper 2003). According to ERP's executive director, Marjorie Mayfield Jackson,[1] the Corps invested "a lot of time and effort" setting up the fund (Jackson 2012b). The trust fund was unique in that it was the first of its kind to offset impacts to a river bottom (similar funds have been set up for wetlands mitigation). The fund also assured that any remediation work would occur within the watershed instead of outside of it, avoiding a problem that often happens in wetlands mitigation (Harper 2004). ERP's very existence—as well as its reputation for quality work—provided the incentive to establish a trust "where the money could go" to facilitate remediation and restoration (Hume 2012). Meanwhile, the fund allowed ERP to press restoration and mitigation work forward, avoiding the long delays that often accompany Corps' projects (Jackson 2012b).

National Oceanic and Atmospheric Administration

Like the Corps, NOAA has a physical presence in Hampton Roads. This has not only resulted in the agency being a tremendous source of volunteers and expertise for ERP, LRN, and NRPA, but NOAA's Marine Operations Center, Atlantic, which is located in Norfolk, also became a model-level "River Star" for ERP within one year after reaching out to the group and improving the facility with native plants, electricity and water-usage efficiencies, and green roofs (Harper 2006). As part of its overall mission, the agency operates various hydrographic survey, oceanographic research, and fisheries survey vessels; the Atlantic vessels are managed by the Atlantic Marine Operations Center. NOAA also has a Chesapeake Bay Office located at Nauticus, the National Maritime Center and Museum in Norfolk, and at VIMS in nearby Gloucester, Virginia (NOAA 2012). NOAA's Chesapeake Bay Office works to protect and restore coastal and marine habitat and partners with the CBP.

The ERP, in particular, has benefited from NOAA's expertise and support. Recently, for example, the group has received a series of large grants from the agency, including $140,000 in September 2012 to support ERP's Learning Barge, and $430,000 to fund a partnership between ERP and CBF to rebuild oyster reefs and wetlands in the Lafayette River, a branch of the Elizabeth, and a wetlands restoration project on the Elizabeth overseen by the City of Norfolk (Harper 2012a). NOAA also participates in the cleanup of the three Superfund sites on the Elizabeth, and developed a geographic information system (GIS) Watershed Database and Map to support restoration work (Elizabeth River Project 2012a).

Perhaps most significant, however, has been NOAA's human contribution. Captain Steven R. Barnum, as the director of NOAA's Office of Coast Survey, served as the nation's chief hydrographer until he retired; he is now assisting with NRPA's John Smith Water Trail mapping project and is involved with its Wetlands Committee (Barnum 2012). As described in chapter 4, Walter Priest, a scientist with NOAA's office in Gloucester who grew up in Norfolk, has participated in the design and construction of more than eighty acres of habitat restoration in over forty separate projects in the Hampton Roads area (W. Priest 2012b). He also served on the board of ERP for about ten years (W. Priest 2012b). Numerous interviewees pointed to Priest as an incredible source of expert information and hands-on assistance (Stedfast 2012; Host 2012; Copeland 2012; Rieger 2012). As Mike Host, environmental programs director for Norfolk Naval Shipyard, explained, "He knew what made these ecosystems tick—he would come in and help with the planning and design of the projects, and be there with his rubber boots on in the mud with everyone else" (Host 2012). Meanwhile, Ray Moses, a retired NOAA admiral who served as ERP's board chairman from 1995 to 2003, is widely credited with providing ERP with critical stature and credibility, creating connections with city, business, and federal leadership (J. Priest 2012; Oliver 2012). Moses also brought in crucial expertise and fellow board members, including Thomas, a retired engineer from USACE, who eventually succeeded Moses to serve as ERP's president in 2003–2004 (Thomas 2012).

The Environmental Protection Agency

As the nation's primary environmental regulatory agency, EPA is well known for its enforcement authority and its "command-and-control" approach. Unquestionably, the waterways in the Hampton Roads region are cleaner than they were prior to the Clean Water Act, when industries freely dumped toxic waste into the Elizabeth River and sewage-treatment controls were lax or nonexistent. In the early 2000s, EPA opposed Maersk Sealand's proposal to build a large commercial port on the Elizabeth River in Portsmouth, ultimately withdrawing its opposition only after Maersk significantly increased its environmental compensation funding (Harper 2003).

More recently, EPA's regulatory presence in Hampton Roads has increased because of a lesser-known but increasingly important provision of the Clean Water Act—Section 303(d), which provides for Total Maximum Daily Loads or "TMDLs." If point source controls are implemented but water-quality standards nevertheless remain unmet, the Clean Water Act requires states to take a second step: they must define the total amount of specific pollution a water body can take (the TMDL) and then distribute the pollution loadings among point source dischargers *and* nonpoint and natural sources. The TMDL must be designed to bring the water body in compliance with the applicable state water-quality standards. States must then work to meet the new TMDL limits, which usually requires point sources to face more stringent permit requirements and raises the possibility for increased state regulation of nonpoint sources (Houck 2002). EPA and the states generally did not take the TMDL provision very seriously for the first twenty-five years of the Clean Water Act's existence until a series of court cases pushed them to do so (Salzman and Thompson 2012). As we describe later in this chapter, the Bay TMDL has raised EPA's profile in Hampton Roads significantly over the past two years, as its requirements are creating new pressures on the region to achieve increased reductions from sewage-treatment plants, develop stronger stormwater controls, and develop better land-use practices to control nonpoint source pollution.

Perhaps less appreciated, however, is EPA's role in providing significant funding and technical support for restoration and education for local watershed groups. As a source of funding and technical support, EPA's role in Hampton Roads looms especially large for the Elizabeth River, given that it remains one of the most polluted rivers in the country. In 2008, for example, ERP received almost $1.2 million from EPA to clean up an area called Money Point, one of the river's most contaminated sections. Meanwhile, millions have been spent assessing the three Superfund sites on the river, and millions more will be spent cleaning them up. EPA has designated an entire team to address remediation on the Elizabeth, partnering directly with DEQ and ERP as part of a "more comprehensive approach" to coordinate the river's remediation work and the overall cleanup approach (US Environmental Protection Agency 2012c).

EPA is also demonstrating a growing understanding of the important role local environmental collaborations have, and will continue to have, in addressing pollution problems (Wayland 2012). As Robert Wayland, director of the Office of Wetlands, Oceans, and Watersheds for EPA in the Clinton and George W. Bush Administrations, observed, "I think we basically decided . . . there is a place for top down and there's a role for the state and that most of the action happens really locally" (Wayland 2012). Beginning in the mid-1990s, the agency started providing support to local watershed groups and encouraged states to do state-level watershed roundtables (Wayland 2012). As part of these roundtables, Wayland recalled there was significant discussion about the importance of finding linkages among groups at different levels, especially to deal with the thornier problem of unregulated, nonpoint source pollution:

> Some of these other problems where there wasn't a regulatory handle and they got interested in watershed management and we got them to talk about the concept of nesting management or pyramid management where you had lots of Nansemond [River] level groups that might be collaborating with the CBF, EPA, Chesapeake Bay Program—so you have this mutually informing, mutually supporting framework of smaller to larger or larger to smaller. (Wayland 2012)

Wayland went on to observe that, to his knowledge, neither the CBP nor EPA ever developed such a network. Such discussions may well have informed the NGO-driven efforts we describe later in this chapter that appear designed, in part, to create greater collaboration among regional, state, and local watershed restoration groups. And the CBP may yet play such a role. According to Jeff Corbin, senior advisor on the Chesapeake for EPA administrator Lisa Jackson, local implementation is increasingly a topic of discussion for the CBP's Executive Council, which consists of the governors of Virginia, Maryland, and Pennsylvania; the EPA administrator; the mayor of the District of Columbia; and the chair of the Chesapeake Bay Commission. Indeed, when these discussions arise, ERP is often mentioned, as it has a reputation at EPA for being an exemplary local restoration group: "ERP is always in the top couple everyone [at the agency] is talking about" (Corbin 2012). Notably, according to Corbin, LRN's exposure is rising as well. Until the CBP collaborates more closely with local watershed groups and appreciates their value, "it's a missed opportunity," in Wayland's view (Wayland 2012).

The Virginia Department of Environmental Quality

The Clean Water Act allows EPA to delegate authority to the states for the implementation and enforcement of federal clean-water goals, including the National Pollutant Discharge Elimination System (NPDES) permits required for point source pollution dischargers. In Virginia, the State Water Control Board (SWCB)—a seven-member citizen board—has primary responsibility for regulatory programs under the Clean Water Act. DEQ staff supports the SWCB's work, and the DEQ director is responsible for administering and enforcing the SWCB's permit programs and regulations. DEQ is headquartered in Richmond, with a regional office in Hampton Roads.

DEQ has been a particularly important source of information, expertise, and technical assistance for ERP, which perhaps is not surprising given the Elizabeth River's legacy as a highly polluted, "industrial" river having numerous point source dischargers. DEQ, for example, was one of the stakeholders ERP brought to the table when it convened its 1993 meeting to develop a restoration plan for the river. DEQ was also a crucial partner in the development of ERP's living trust fund and a cofunder with NOAA for a large voluntary wetlands restoration project (Jackson 2006; Harper 2012a). DEQ also partners with EPA's Elizabeth River remediation team for Superfund sites on the river (US Environmental Protection Agency 2012c). Significantly, ERP engaged Frank Daniel, the regional director for DEQ for the Hampton Roads area, early on. Daniel (2012) described how, in the early 1990s, Marjorie Mayfield Jackson, ERP's executive director, approached him "to help them get started and if possible help give them insights into grant money through the state and at the federal level for the EPA." Daniel worked closely with ERP while he was regional director and joined ERP's board after he left DEQ. Daniel has been a key advisor to ERP, providing crucial advice ranging from the political—how to work with federal and state policy makers—to the technical—how to conduct effective water-quality monitoring and how to evaluate the quality of engineering-firm proposals for restoration work (Daniel 2012; Moses 2012).

DEQ water-quality data has, of course, been a critical part of the information necessary for all of the groups we studied to monitor and evaluate the water quality for their respective rivers, although funding for water-quality monitoring unfortunately has been reduced in recent years. Nick Worth, chair of NRPA's Water Quality Committee, described how incredibly useful DEQ's monitoring records were to NRPA and his committee's voluntary water quality monitoring work (Worth 2012b). Because of a lack of funding, however, DEQ's monitoring on the Nansemond stopped and, according to Worth, the agency was "anxious" to work with NRPA, encouraging the group to participate in its Citizens Water Quality Monitoring Program (Worth 2012b).

The Virginia Department of Conservation and Recreation

In addition to its traditional role of managing the state's parks and recreational areas, the Virginia DCR manages soil and water conservation, oversees stormwater regulation, implements nonpoint source pollution programs, and implements Chesapeake Bay Local Assistance programs. Consequently, the agency plays a large role in Hampton Roads as a source of technical expertise and regulatory guidance for the grassroots collaboratives we studied. DCR has been especially involved with ERP and LRN with respect to providing technical assistance for nutrient management and wetlands restoration (Hill 2012; T. Herbert 2012).

DCR's presence in the region as a policy convener has arguably elevated over the past few years as Virginia has responded to the Bay TMDL. Once EPA established pollution limits for each state under the Bay TMDL, the states were required to develop "Watershed Implementation Plans" ("WIPs"), where they identify the measures they plan to implement in order to achieve the Bay TMDL's pollution-reduction requirements and restoration goals (US Environmental Protection Agency 2010a). The second phase of this process—known as the "Phase II WIP"—was completed in 2012 and required a great deal of information from local governments. James Davis-Martin, DCR's Chesapeake Bay TMDL coordinator, has been responsible for coordinating and conducting numerous "WIP town hall" meetings around the state for the past two years—including several meetings with the HRPDC that involved ERP and LRN (Davis-Martin 2012).

The Chesapeake Bay Program

The Chesapeake Bay's importance to the region is evidenced by the fact that a regional federal-state partnership known as the CBP exists solely to restore and preserve the Bay's waters and watershed. Described as "a partnership that directs and conducts the restoration of the bay" on its website (Chesapeake Bay Program 2012a), the CBP is a bureaucracy created in 1983 by Section 117 of the Clean Water Act and is also a creature of voluntary agreements made by its federal and state partners, making it a good example of a government-sponsored voluntary environmental program (VEP).[2] The CBP consists of programs, initiatives, and organizational policies that reflect a series of formal, nonbinding agreements (discussed in more detail later in this chapter) that Bay jurisdictions have entered into over the course of the CBP's thirty-year existence. Renowned for its science but lacking in regulatory power, restoration action requires CBP partners to agree to develop and implement concrete regulatory and land-use-management programs—a notoriously slow and complex process for the CBP's federal, state, and local stakeholders that manifests itself by means of the CBP's organizational structure (Wiersema 2008). The Executive Council—predominantly governors in the watershed—meets annually and sets goals for the CBP, but much of the substantive decision making and agenda setting is delegated to the Principals' Staff Committee (PSC), which primarily consists of high-level, career environmental regulators and resource managers representing various federal and state agencies. Meanwhile, the PSC's work is informed by the findings and recommendations of numerous committees and subcommittees. Several individuals we interviewed for our case studies participate or have participated on some of these committees. Debbie Ritter, for example, a member of the City of Chesapeake's City Council and an ERP board member, serves on the CBP's Local Government Advisory Committee. Christy Everett, executive director of the CBF's office in Hampton Roads, serves on the Citizens Advisory Committee. Walter Priest, former board member at ERP and NOAA wetlands expert, serves on the Wetlands Action Team.

The Hampton Roads Planning District Commission

The HRPDC is a regional organization that includes and represents sixteen local governments in the Hampton Roads area. The Virginia General Assembly created Planning District Commissions (PDCs) in 1968 when it passed the "Regional Cooperation Act." Under Section 15.2-4207 of the Virginia Code, the purpose of PDCs is "to encourage and facilitate local government cooperation and state-local cooperation in addressing on a regional basis problems of greater than local significance." The HRPDC not only provides a forum for local and elected officials and administrators to deliberate, but it also provides strategic-planning services and technical expertise to member governments "on local and regional issues pertaining to Economics, Physical and Environmental Planning, Emergency Management, and Transportation" (HRPDC 2012a). The HRPDC's Physical and Environmental Planning Department particularly emphasizes water-resource management, with stormwater management and the Bay TMDL becoming important areas of focus in recent years (Tribo and Katchmark 2012).

The HRPDC interacts with local watershed collaboratives in several ways. When the USACE conducted two major evaluations of restoration projects in Hampton Roads—one in the Elizabeth River and the other in the Lynnhaven—the HRPDC served as a coordinator for the projects. According to Whitney Katchmark, HRPDC's Principal Water Resources Planner, "LRN and ERP were both very active . . . in terms of giving input on sites and how the Corps should approach property owners or general citizens" (Tribo and Katchmark 2012). Other projects have included developing TMDL-implementation plans, working with localities, watershed groups, and DEQ. While all of this sounds very positive, at least one local environmentalist we spoke with did not believe the nonprofit environmental community was sufficiently "at the table" with the current TMDL process. As the individual put it, "We can come and observe but we don't have a voice."

The HRPDC played and continues to play a significant role with respect to the Bay TMDL. Initially, the HRPDC's stance toward the Bay TMDL was hostile—the group considered filing a lawsuit against EPA to challenge the Bay TMDL's pollution-reduction model and expense, estimating that implementing the TMDL requirements may cost "each household [in the region] between $350 and $1,000 a year" (Harper 2011a). The commission ultimately decided not to sue, after publicly crediting EPA's senior advisor on the Bay, Jeff Corbin, for entering into an open dialogue with the Commission and making an effort to be more flexible (Harper 2011a). Support and pressure from important community members associated with local environmental groups also appear to have played a role (Tribo and Katchmark 2012). Everett, executive director of CBF's Hampton Roads Office, described the commission's decision as a "trifecta" that involved the commission's willingness to engage, grassroots pressure organized by CBF and other groups, and "grasstops" pressure from leaders in the community associated with or supportive of LRN, ERP, and NRPA (Everett 2012). For example, Andy Fine, the politically connected cochairman of LRN, coauthored with Louis Ryan, a well-known retired lawyer and donor of a $25,000 challenge grant to ERP, an opinion piece in support of the Bay-wide TMDL that ran in *The Virginian-Pilot* (Ryan and Fine 2011).

While the HRPDC was considering legal action, it simultaneously had to respond to requests from EPA and the state for coordination, planning, and information as the Bay TMDL process proceeded. For example, when Virginia was tasked by EPA to develop the "Phase II Watershed Implementation Plan" and develop pollution-reduction goals and plans for the local level, state officials requested PDCs to coordinate local stakeholder meetings and gather data. HRPDC created local government working groups and a steering committee that included government representatives from each locality, professionals from federal and state agencies, the agricultural community, and environmental groups, including CBF, LRN, and ERP (HRPDC 2012a). Whitney Katchmark at HRPDC described how critical it was to have participation from these groups because they were able to convince local government officials that many private-property owners were interested in taking action:

> [At] the initial meetings most of the local government representatives were very skeptical. They didn't think anybody would participate in such a program and they didn't think people would want to do retrofits on their private properties and they didn't really want to ask them to. I think at that meeting, it was probably the first time that a lot of the . . . CBF and LRN were there and they spoke up and were like, "We got people to do it or people have asked us how they can do it." There are people out there that want to do stuff and need guidance. They want to do something and if you gave them an incentive that might push them over the edge. There are willing participants out

> there. I think that was helpful for the local government to hear that. That kind of led into this bigger discussion and we actually ended up doing this project that was supposed to be part of the Bay TMDL and Phase II project. (Katchmark and Tribo 2012)

In addition to providing support for local government as they determine how to invest the infrastructure and programs necessary to improve water quality, Katchmark also observed that local watershed collaboratives are important for "getting people involved in the process":

> Local governments, as hard as they try, just are not very good at it. Government in general is not very good at engaging its citizens. It seems even as we have gone through this TMDL process, I have heard people ask how the environmental groups got in the area. They would like one. Local governments see the opportunity for it. I think they still can be skeptical when things come up because it can be adversarial at times. I think they see the benefits of a LRN and doing what they do. (Katchmark and Tribo 2012)

In this way, local watershed collaboratives appear to play a critical role in both legitimizing government action and engaging citizens. Involving more people in the TMDL process appears to have had the unintended benefit of demonstrating to both local government and individual citizens the potential benefits to the community that an engaged grassroots collaborative willing to cooperate with various partners can bring.

Key Regional and State Environmental Groups

When it comes to large environmental groups, the CBF, an environmental nonprofit that has advocated for the Bay for more than forty years, is, by far, the most important regional watershed restoration group in Hampton Roads. The Virginia Conservation Network (VCN)—a statewide network of environmental groups that includes as members two of the three watershed groups we studied—also plays a role and facilitates local interaction with the Choose Clean Water Coalition (CCWC), an emerging advocacy group in the region. While not an environmental group, the Virginia Environmental Endowment, a foundation, has been an important supporter for each of the collaboratives we studied.

The Chesapeake Bay Foundation

Founded in 1967, the CBF is the region's largest environmental group dedicated to restoring the Chesapeake Bay, having more than two hundred thousand members (Chesapeake Bay Foundation 2012a). CBF describes itself as "[s]aving the Bay through education, advocacy, litigation, and restoration" (Chesapeake Bay Foundation 2012a). From lobbying Congress to appearing before state agency rulemaking bodies to conducting local oyster gardening projects, CBF works on behalf of Bay restoration on the national, regional, state, and local levels. While the group has been criticized by long-time Bay analysts such as Howard Ernst for not being aggressive enough in its litigation or advocacy strategies, it is perceived by many in the Hampton Roads area as having the most forceful and adversarial approach in the region and Virginia (Daniel 2012; J. Priest 2012; Smith 2012a; Linkenhoker 2012). It also has served as a training ground for many local and state environmentalists, who have either started other groups or have joined other organizations (Stiles 2012).

Headquartered in Annapolis, Maryland, CBF has offices in Washington, DC; Harrisburg, Pennsylvania; and Richmond, Virginia. CBF's Hampton Roads office, located in Norfolk, Virginia, was the organization's only office not located in either the United States capitol or a state capitol until it opened an office in Easton, Maryland, in June 2012. Mike Kensler, prior to becoming one of the original members of the "kitchen table" that led to the ERP, opened CBF's Hampton Roads office in 1990 (Kensler 2012). According to Kensler, CBF located in Hampton Roads because "the mouth of the Bay" made it "important for CBF to have a presence" (Kensler 2012).

At least two grassroots efforts appear to have paved the way for CBF to establish an office in Hampton Roads (Kensler 2012; McCarthy 2012). In the mid-1970s, Citizens Against Refinery Effects (CARE) formed in Norfolk to stop a refinery from being built in Portsmouth (Kensler 2012; Everett 2012). Lacking the time to create a nonprofit entity to fundraise, CARE turned to CBF, which agreed to affiliate with the group (Hines 2002). According to CARE leaders, CBF's credibility helped CARE raise $100,000, meet with important political decision makers, and defeat the refinery. After what was described as an "ugly fight," the group—known as the BayCare Chapter of CBF—was successful (Hines 2002). CBF began considering opening an office in Hampton Roads, building from the BayCare Chapter (Kensler 2012). Sharon Adams, a member of ERP's initial "kitchen table" meetings, noted also that there was more interest at that time on CBF's part in "talking about rivers and tributaries" (Adams 2012b). In addition to CARE's success in Hampton Roads, Adams pointed to "Clean the Bay Day," a successful grassroots effort organized by another ERP founder, Robert Dean. According to Adams, all of this indicated to CBF that the area could and would support broader environmental engagement (as well as become a good source of fundraising), prompting CBF to commit staff resources in the Hampton Roads region. Scott Harper, longtime environmental reporter for the region's newspaper, the *Virginian-Pilot*, provided a similar assessment (Harper 2012b).

With a staff of ten that includes a senior scientist, grassroots coordinator, and two oyster scientists and specialists, CBF's current presence in Hampton Roads is strong. The organization is widely known and recently has been particularly active in informing the Bay TMDL process. It also has recently partnered with ERP, the City of Norfolk, and NOAA to restore the Lafayette River, a branch of the Elizabeth River, with its "oyster team" playing an important role (Everett 2012). In October 2012, CBF announced it would build a "state-of-the-art environmental education and community center" in Virginia Beach, Virginia, on a parcel known as Pleasure House Point, which CBF purchased from the Trust For Public Land and was acquired by the City of Virginia Beach for conservation. LRN will be sharing space with CBF at the new center.

All of the environmental groups we interviewed pointed to CBF as an important voice for watershed restoration policy—and, indeed, often described CBF as fulfilling an important advocacy niche on a statewide and regional level that usually complements their organization's local advocacy and restoration efforts (Johnson 2012; Forget 2012a). Similarly, CBF is an important partner for LRN and ERP for oyster gardening and restoration (Forget 2012b). CBF has also been a crucial model and supporter for the groups we studied: its Hampton Roads executive director, Kensler, was one of the four original kitchen-table founders of ERP; Fine describes how he and Harry Lester "immediately called the CBF and Will Baker," CBF's president, when they decided to start LRN; and NRPA's first consultant, Wilson Browning, came to CBF for institutional information, a regulatory and policy overview, and an understanding of the scientific research applicable to the Nansemond River (Fine 2012; Everett 2012). Dick Barry similarly described how CBF agreed to be a financial "pass through" for donors as NRPA worked to establish 501(c)(3) status (Barry 2012).[3] While very well-respected as a group having strong scientific, policy, and organizational expertise, some distrust toward CBF surfaced occasionally in some of our interviews in off-the-record comments not for attribution. For example, some interviewees expressed frustration that CBF gets too much credit for ideas it did not initiate (Clean the Bay Day, for example) or that the public may consider projects implemented by the grassroots collaboratives to be CBF's work. Certainly, as Wayland pointed out, there's always going to be "a little bit of organizational rivalry that goes on and the CBF wants to be seen as *the* watershed organization" in the region (Wayland 2012).

Because CBF's approach is perceived as sometimes more adversarial than collaborative, some of the members of the local watershed groups we interviewed indicated that, at times, they worked to distinguish themselves from CBF by taking a less confrontational approach (Moses 2012; Smith 2012b). CBF, for its part, appreciates the reasons for these groups in taking a less adversarial tack—and, indeed, appears to value it and endeavors to take such a tack itself in Hampton Roads when it can. As Everett, CBF Hampton Roads director, explained, "You need [a polluter] to be scared, but you need them too to feel the support and interest of the community to just improve and focus on water quality" (Everett 2012). In Everett's view, grassroots support designed to help local officials, businesses, and citizens make the investments necessary to improve water quality is essential—and groups such as ERP, LRN, and NRPA are well positioned to fulfill this role. In regional discussions about implementing the Bay TMDL, Everett explained that "without LRN having the power to say 'I'm standing here speaking on behalf of 3,000 members and we have credibility, don't delay,' [the discussions] would not have been as effective" (Everett 2012). Instead of threatening legal action, "groups are saying we can do [it] together with people power and work it out on a civil basis by talking to our elected officials and [creating] behavior change" (Everett 2012). Moreover, Everett describes local grassroots collaboratives as "neutral territories" that "resonate with trust" and help develop relationships that make people more inclined to change their behavior and donate their time and resources to restoration (Everett 2012).

Everett's observations were echoed by many others. Mayfield Jackson, ERP's executive director, described CBF's policy and advocacy work as "an example of complementing the collaborative approach" (Jackson 2012b). Karen Forget, LRN's executive director, perceives CBF as conducting similar work as LRN but on a different scale. "I don't think it was ever really a matter of stepping on somebody's toes. CBF is a big regional organization that does a lot of the same kind of stuff that we do. We do it in a more concentrated level. They do it on a regional level" (Forget 2012a). DCR employees, including Davis-Martin, DCR's Chesapeake Bay TMDL coordinator; Noah Hill, regional manager for Hampton Roads; and Todd Herbert, watershed field coordinator, each perceived that local watershed collaboratives have a strong ability to establish a trusting and working relationship with the localities in which they are located (Davis-Martin 2012; Hill 2012; T. Herbert 2012). They also gave CBF's more aggressive reputation credit for allowing the local groups to take a more collaborative approach, suggesting CBF "freed up" the local groups to be more cooperative. While they distinguished CBF's regional approach from the very local approach conducted by groups such as LRN, they also distinguished CBF's regional office from its larger organizational identity, observing that the Hampton Roads CBF office appreciates the role of local grassroots collaboratives and appears to work with them whenever possible. Moreover, CBF has tremendous resources, both to make the lobbying case for increased federal and state funds to come to Hampton Roads and to provide technical assistance. As Hill observed when discussing an oyster restoration project in the Lynnhaven River, CBF was "able to give us resources that weren't available to the other non-profits" (Hill 2012).

The Virginia Conservation Network and the Choose Clean Water Coalition

An affiliate of the National Wildlife Federation, the VCN describes itself as "a coalition of nonprofit and community groups that work together for a cleaner, healthier environment" (Virginia Conservation Newtork 2012). VCN conducts public-policy research, advocacy, and provides "capacity building" for its member organizations. Nathan Lott, VCN's executive director, explains that the largest interest groups represented in VCN consist of watershed groups, energy groups, and wildlife or land trust groups. VCN has more than one hundred participating member organizations, including ERP and LRN. VCN is affiliated with the CCWC, which was formed in 2008 with a mission to "serve as a strong, united, effective advocate for restoring the thousands of streams and rivers flowing to the Chesapeake Bay by coordinating policy, message action and accountability for clean water at the federal, state and local levels" (Choose Clean Water Coalition 2012). VCN is the state lead for CCWC for following and working to inform the Bay TMDL in Virginia (Marx 2012).

Of the three groups we studied, LRN is the most oriented toward influencing public policy and thus is most engaged with state and regional coalition work (Stiles 2012). Forget (2012a), LRN's executive director, described how VCN and CCWC play a crucial role in keeping her group informed about state and federal policy:

> One of the challenges for a smaller local group like ours that is really very locally focused is just to keep up with the issues at a state and federal level. . . . [VCN] really serves a very valuable function for smaller groups like ours gathered throughout the state that have a more local or specialized focus. They can be our ears and eyes in Richmond and keep us informed of what is going on in Richmond. They do briefings and facilitate conservation lobbying day with the gardening clubs. They serve a really important function for groups like ours.

Forget also found CCWC's book on federal policy to be very helpful, noting that "[w]e take copies to all of our delegates and senators and we give copies to our city council members, too" (Forget 2012a). She went on to explain that "[t]hey develop these two page briefings on all of the major topics that have to do with conservation. Some of them are more pertinent to us directly" (Forget 2012a). LRN's public-policy chairperson is especially active and participates in VCN and CCWC conference calls on LRN's behalf. Because keeping up with emerging public policy and legislation is quite a time investment, Forget laughingly explained, "I am really happy that she is retired and wants to do those calls" (Forget 2012a).

For their part, both VCN and CCWC also benefit from having local grassroots as participating members. As Lott explained, local groups are "going to have the best relationships with their legislators oftentimes" (Lott 2012). Peter Marx of the CCWC agreed, describing how LRN was a great asset in expressing to Virginia Beach's congressman, Representative Scott Rigell (R-VA), how important certain Farm Bill programs were for local water quality (Marx 2012). Legislators "listen to locals," Marx explained. Moreover, given that the Bay TMDL is having such an impact on local governments, the participation of local grassroots groups in supporting the TMDL effort is likely to become only more important (Marx 2012). Davis-Martin (2012) at DCR observed that "groups like Virginia Conservation Network" are "surprisingly sophisticated" in how they "bring more political pressure on behalf of those smaller organizations."

The Virginia Environmental Endowment

The Virginia Environmental Endowment (VEE) is the most prominent foundation in Virginia focused on funding conservation and educational projects designed to improve the environment. VEE was founded in 1977 after Judge Robert Merhige of the US District Court in Richmond imposed a criminal fine of $13.24 million against Allied Chemical Corp. for discharging two hundred thousand pounds of Kepone—an insecticide—into the James River (Davison et al. 1997). Judge Merhige, dismayed that the fine would go directly to the federal government, agreed to reduce the fine to $5 million if Allied would set aside $8 million to fund environmental work in Virginia. Since its founding, the endowment has awarded approximately $27 million as part of more than 1,200 grants and has been led by Gerald McCarthy—an individual who was regularly cited by interviewees as a critical supporter of their projects. McCarthy has a reputation for funding projects in their initial stages, and this was evidenced by his early support for ERP. He fondly recalled how the four original founders of ERP—Adams, Mayfield Jackson, Dean, and Kensler—came to visit him in Richmond:

> I guess Sharon [Adams] called me up—and said, "I got a great idea for you. We're forming a group and we need your help to get it off the ground." This is what later became the ERP. So, they came—the four of them—to see me in Richmond, and we sat around the table in my office and had quite a discussion, and I was so tickled—these people who claimed that they'd dreamed this idea up while drinking beer and eating pizza a recent night beforehand, waking up the next morning realizing they'd actually agreed to restore the Elizabeth River and if they were serious about it, they better come up with some money and a plan—so they came to see me and spent an hour explaining how and what they wanted to do and I was really intrigued—and saying, "Oh man, I wonder how much money they want"—so finally I asked how much—and Sharon says, "$1,375." And I said, "DONE!" Just like that. "Just get me some paperwork." So we made that grant fast before they changed their minds. It's grown into millions—and of course we've given much more money in the twenty years since then, but that's what started it. (McCarthy 2012)

VEE has continued to support ERP over the years with grants significantly larger in amount. The initial grant, however, remains fresh in the minds of McCarthy and ERP staff and supporters (McCarthy 2012; Adams 2012b; Jackson 2012a). Howard Copeland, a former member of the Virginia House of Delegates and ERP Board Member in the 1990s, described McCarthy as "lifeblood at the outset" (Copeland 2012). Two founders of LRN—Lester and Fine—likewise sought out McCarthy for advice "because they had a new river group they were starting and wanted to pick my brain about how to go about doing it" (McCarthy 2012). McCarthy recalled providing LRN with a "mini-grant of $5,000" and "they were off and running." NRPA also received a mini-grant from VEE in its early stages—as well as advice from McCarthy (Taraski 2012a; McCarthy 2012). As different as the groups we studied are, McCarthy described that each of them had something in common in the beginning: the leadership are "all initiators and all willing to commit to getting something particular done and they persist until they see results" (McCarthy 2012). For its part, VEE represents how both start-up funding and advice from grant-makers can greatly assist local grassroots organizations in gaining credibility and confidence to proceed with developing and implementing their organizational and environmental goals.

CHESAPEAKE BAY ENVIRONMENTAL OVERVIEW: SAVE THE BAY MEETS RIVER RESTORATION

Regional collaboration has been a hallmark of Bay restoration efforts since the mid-1980s, with the CBF playing the primary environmental advocacy role. As we describe below, the regional partnership that has dominated Bay restoration work has evolved over time, as new information about the sources of pollution and the mechanisms to address it have become better understood and as regional restoration goals have been missed. Meanwhile, as the primary regional collaborative effort, the CBP, faltered, new grassroots collaborations focused on local waterbodies—groups like ERP, LRN, and NRPA—appeared. A more complicated approach to Bay restoration efforts may well be emerging—one that requires an appreciation of the possible limits of regional collaboration as well as the potential benefits of grassroots collaboration on a local level to complement "top down" requirements.

Bay Restoration: An Evolving Regional Approach

While residents living in the watershed have been concerned about pollution in the Chesapeake Bay for more than a century, the concern was primarily local (Boesch, Brinsfield, and Magnien 2001). With the exception of overharvesting fisheries, the Bay's woes were largely linked to industrial pollution from sources such as power and chemical plants, municipal pollution from sewage-treatment plants, and channel dredging—all problems a concerned citizen could identify by pointing to a specific name and location. This localized understanding of the causes of Bay degradation changed, however, in the 1980s, when scientists began to document how eutrophication—the overenrichment of waters by organic matter that depletes oxygen in the water—was occurring throughout the Bay from a variety of nontraditional pollution sources, including nutrient runoff from agriculture and development (Boesch, Brinsfield, and Magnien 2001). A bigger and more diffuse picture of the Bay's problems thus emerged, and with it came a regional, Bay-wide restoration approach.

The Bay-wide restoration focus began in earnest in 1983, after a $27 million study sponsored by US Senator Charles "Mac" Mathias (R-MD) explained how excess nutrient pollution was causing the Bay's rapid decline. Soon after the study was released, the first of three "Bay Agreements" was signed by the governors of Maryland, Virginia, and Pennsylvania; the mayor of the District of Columbia; and the administrator of EPA (Chesapeake Bay Agreement 1983). Congress then took the agreement a step further by incorporating it into the Clean Water Act, creating the CBP and designating the agreement's signatories as the "Executive Council" that "directs" the Bay Program (33 U.S.C. § 1267(a)(4)). The CBP is the oldest regional watershed partnership in the country.

The three agreements that CBP partners have signed over the past three decades have profoundly shaped the CBP's approach and focus. Signed in 1983, the first Bay Agreement was modest in scope—a one-page document that created a group of state "cabinet designees" to oversee the coordination of Bay restoration efforts. A short document, the 1983 Agreement nevertheless set the tone for the voluntary (and oft-criticized) culture of the Bay Program for decades to come, stating that "a cooperative approach is needed . . . to fully address the extent, complexity, and sources of pollutants entering the Bay" (Chesapeake Bay Agreement 1983). It did not include legally binding commitments.

A second and more robust Bay Agreement followed in 1987 that created far more substantive and detailed—although still nonbinding—policy goals, including an agreement to reduce nitrogen and phosphorus pollution by 40 percent by 2000 (Chesapeake Bay Agreement 1987). It also replaced the cabinet secretaries with the governors of Virginia, Maryland, and Pennsylvania and included both the EPA administrator and the chair of the Chesapeake Bay Commission. In 1992, the 1987 Agreement was amended, reaffirming the 40 percent reduction goal and expanding the CBP's focus beyond the basin itself by emphasizing the importance of addressing nonpoint source pollution in the tributaries draining into the Bay (Chesapeake Bay Agreement 1992).

The CBP missed the 2000 deadline of reducing nutrient pollution by 40 percent, and a third agreement—Chesapeake 2000—was signed. Chesapeake 2000 was a comprehensive and ambitious document that acknowledged, for the first time, the importance of local and individual efforts, setting as one of its goals promoting "individual stewardship" and assisting "individuals, community-based organizations, businesses, local governments and schools to undertake initiatives to achieve the goals and commitments of this agreement" (Chesapeake Bay Agreement 2000). Significantly, it also included a goal to remove the Bay and the tidal portions of its tributaries from the Clean Water Act's impairment list by 2010—a goal set, in large part, because of two consent decrees resulting from court cases in Virginia and the District of Columbia (Federal Register 2009).[4]

By 2008, however, it was clear that the CBP would fall far short of meeting this goal, and CBP partners discussed openly the problems that the program's lack of success created for its credibility with the public. The CBP's failure also led to President Obama issuing, in 2009, an Executive Order on Chesapeake Bay Protection and Restoration—the first Executive Order of its kind to focus on a specific watershed and which required EPA to "examine how to make full use of its authorities under the Clean Water Act to protect and restore the Chesapeake Bay and its tributary waters" (Executive Order 2010). EPA responded in 2011 with the "Bay Total Maximum Daily Load" (Bay TMDL), which includes new pollution limits for each of the ninety-two "impaired water quality segments" flowing into the Bay and constitutes the nation's first regional TMDL (Federal Register 2011). The Bay TMDL is striking for several reasons: EPA takes a large role in both assuming responsibility and setting "expectations" to the states for action in a series of detailed letters; President Obama tasks federal agencies with responsibilities for cleanup; and EPA administrator Jackson is instructed to use Clean Water Act authority to the "fullest extent" to improve Bay water quality (Houck 2011). Regional pollution-reduction loads also were set that would have to be divided further by the states in WIPs on a subwatershed and water-segment level (Houck 2011). In a new tactic, EPA made the WIPs "cornerstone of the accountability framework" for ensuring real gains are made and set two-year milestones for evaluating progress (Houck 2011).

We review this history for several reasons. The CBP's "cooperative partnership" and subsequent missed deadlines have generated harsh criticism of collaboration as an effective approach to Bay restoration. Professor Howard Ernst at the US Naval Academy—a long-time observer of Bay-restoration efforts—is particularly pointed in his criticism of the CBP and of collaboration to achieve meaningful environmental goals. Ernst rightly observes that the CBP is a "regulatory agency with no regulatory powers; created to enforce Bay Agreements that are not legally enforceable" (Ernst 2010). And we have no real quarrel with Ernst's assessment that, consequently, the CBP does little more than nudge its state partners forward instead of forcing them to invest in expensive stormwater controls and sewage upgrades or to regulate agriculture, the Bay's largest source of pollution.

But the failures of a government-organized regional collaboration with a quasiregulatory veneer should not cloud the potential for citizen-led, grassroots collaborations to generate the necessary private and local government action necessary to improve local water quality. Unlike the collaboratives we studied, the CBP is primarily an agency-based partnership of institutional actors (Heikkila and Gerlak 2005; Margerum 2008). Moreover, many of these actors have varying levels of commitment to the CBP's goals because they have little to gain from their efforts—and much to lose by alienating important interest groups such as farmers. For example, Pennsylvania significantly contributes to Bay degradation because of nutrient pollution from its farms that drain to the Bay via the Susquehanna River, but the state itself does not border the Bay and receives little to no economic, cultural, or recreational benefit from it. As such, a crucial component for effective collaboration—mutual benefit with each partner having a stake in participating—is glaringly missing from a key partner of the CBP. As Barbara Gray observes, "Unless some consensus is reached about who has a legitimate stake in an issue and what that joint issue is, further attempts at collaboration will be thwarted" (Gray 1985).

Similarly, BIMBY (Because It's My Back Yard) and its underlying connection to place is also missing for states such as Pennsylvania when it comes to the Chesapeake Bay—not to mention some of the residents of Maryland and Virginia who live hundreds of miles from the Bay themselves. As we observed in chapter 1, community commonality doesn't exist on a regional level in the Bay either. The CBP's scale may simply be too large for genuine collaboration to be effective. Regardless, in our view, the CBP's failure to achieve its restoration goals ignores its success in developing good science and supporting critical restoration expertise and reflects more about the program's limitations and the inevitable politics involved in agency-based partnerships than the effectiveness of collaboration itself as a way to solve local problems. Arguably, the complex scientific modeling underlying Bay TMDL could not have been conceived and conducted without the science and expertise that was developed as part of the CBP's work (Corbin 2012). Indeed, the development of the Bay TMDL may provide just the regulatory pressure that Ernst contends the CBP needs for its partners to make substantive environmental gains. For example, in the two years after the Bay TMDL effort began, the Maryland General Assembly required new regulations for proposed developments using septic systems, citing specifically the Bay TMDL as the impetus for the legislation (Maryland Sustainable Growth and Preservation Act 2012). Meanwhile, EPA withheld funds from Virginia for the first time until the state agreed to revise stormwater permits (known as "MS4 permits") in a stricter fashion consistent with the Bay TMDL and on a tighter timeline (Harper 2012c).

The CBP's history also reveals an increasing awareness among Bay partners of the complexity of the Bay's ecosystem and a greater need for focusing on local water-quality improvements if the Bay, as a whole, is ever to be restored. Only in 2000 was the importance of local and individual efforts officially acknowledged. While prodded in part by the looming possibility of a series of TMDLs being set for Bay segments and tributaries, the CBP's evolution may also represent a growing emphasis on local implementation from EPA and the states as well as increased engagement by local groups. Davis-Martin's recent experience coordinating Bay TMDL efforts in Virginia is instructive. Given how complicated and jargon-filled the Bay-wide TMDL process has been, Davis-Martin observed that many of the citizens he encountered in the Hampton Roads meetings were highly informed about the TMDL, the scientific modeling underlying the TMDL, and the WIP-development process generally (Davis-Martin 2012). He credited the educational work that groups such as LRN and ERP (and CBF) have conducted to an increased understanding of the issues involved with addressing nonpoint source pollution, the Bay TMDL's focus. Because controlling nonpoint source pollution will require increased investment by local government as well as increased private action, Davis-Martin and his colleague Herbert also predict that groups such as ERP, LRN, and NRPA will become even more important partners for local governments. Many of the best management practices or "BMPs" necessary to meet Bay TMDL requirements are residential, Davis-Martin explained—and the "social networking" facilitated by local watershed groups will be critical for the BMP implementation that local governments want to see happen. Herbert (2012) observed:

> [When] local governments have to deal with the reality of having to make some implementation on the ground happen . . . I think it's the localities [that] are going to be using non-profits for their community outreach, for their networking, the networking that's going on as a way to get to the residents because these non-profits go out and they have these activities and cities do not have the time or the resources to reach that many people.

Corbin, EPA's special advisor to the administrator on the Chesapeake Bay, had a similar view and predicts local watershed restoration groups will play an increasingly larger role as the states work to meet the pollution-reduction deadlines established under the Bay TMDL:

> That is where the implementation is going to happen. They are going to have to be the people the states are turning to. Here is the list that has to be done. How are we going to get it done? They are going to have to have those local level discussions. It's not just the policymakers. You have to have those groups at the table. Like it or not, but I think they are going to have a bigger role in this in the next ten to twelve years than they ever have. (Corbin 2012)

In sum, Bay restoration efforts appear to be moving toward an era where local action by grassroots collaboratives will become increasingly important. Much of nonpoint source pollution caused by various and diffuse sources can only be controlled by individual behavior change—a far different challenge than presented by discrete point sources, which have been much more successfully controlled under the Clean Water Act's command-and-control scheme. TMDLs are a regulatory mechanism under the act designed to begin to address the problem, but they are notoriously frustrating legal and policy tools because they are complicated to establish and the act does not expressly call for their implementation (Houck 2002). Many TMDLs now exist, but few are actually implemented. Moreover, because nonpoint sources such as agriculture are generally unregulated, meeting TMDLs requires voluntary action by individuals and nonpoint sources unless states and local governments pass new legislation or implement new policies; getting results through the TMDL process is thus profoundly difficult—perhaps impossible—without local grassroots support (Wayland 2012).

Because of the Bay TMDL and because of local interest in improving water quality generally, local governments controlling stormwater and individual property owners changing land-use patterns are the next frontier for improving water quality in Hampton Roads. We predict grassroots watershed collaborations—with their laser-like focus on improving a specific, beloved river important to their community's history, environment, and economy—will be critical to informing and supporting these efforts. In the next section, we briefly review the Lynnhaven bacteria TMDL, a rare success for the TMDL program that demonstrates the value of strong local collaboration.

Meeting Water Quality Goals: Collaborative Success in the Lynnhaven

The Lynnhaven River is home to one of the few examples in the nation of successful TMDL implementation—one that EPA highlights as an example of a "TMDL at Work" and environmentalists point to as a success (US Environmental Protection Agency 2012d; Everett 2012). While the Lynnhaven River TMDL has many lessons for Bay TMDL implementation specifically, it also holds many lessons for local government–nonprofit partnerships generally.

The Lynnhaven, once famous for its oysters, became so contaminated with fecal coliform bacteria that most of the river was closed to shellfishing. In 2005, 1 percent of the river met water-quality standards for bacteria. In 2006, the Virginia DEQ developed a TMDL for bacteria for the Lynnhaven and an implementation plan, which included input and commitments from the City of Virginia Beach and strong participation from LRN (US Environmental Protection Agency 2012d). The plan focused on human and pet sources of fecal coliform, identifying activities such as sanitary sewer improvements, stormwater upgrades, and pet-waste programs. Forget, LRN's executive director, credits the bacteria TMDL for spurring the city to take action: "The bacteria TMDL for the Lynnhaven was really one thing that's motivated a lot of the actions that the city has done—because they're required to do it" (Forget 2012a). More importantly, according to Forget, the bacteria TMDL did not end up as "one more plan on the shelf" because the "the city recognized that it was in their best interest to try and clean up the waterway [because] there was some significant kind of pressure from the community to try and do so" (Forget 2012a).

According to EPA, both ongoing and newly implemented activities identified in the TMDL's implementation plan "played a significant role in reducing fecal coliform levels and restoring the health of shellfish areas" (US Environmental Protection Agency 2012d). By 2007, 29 percent of the river was deemed safe for shellfish consumption (US Environmental Protection Agency 2012d). By 2012, that figure had reached 42 percent, although it subsequently declined that same year to 40 percent (Harper 2012d). Forget (2012a) credits the implementation of many concrete activities in partnership with the city as a crucial part of the bacteria TMDL's success. She also identified the city's willingness to allow LRN to conduct the TMDL's community-education requirement as being both beneficial to the nonprofit and to the city—LRN had a concrete platform for action that was supported by the city, while the city received public-outreach efforts conducted by LRN that were cheaper and more effective than the city could do itself (Forget 2012a). Robert Johnston, director of Public Works for the City of Virginia Beach, agrees with this assessment (Johnston 2012).

Everett, executive director of CBF's Hampton Roads Office, praises LRN for providing critical grassroots support to the City of Virginia Beach so that the city was able to take strong action to implement the plan and meet the bacteria TMDL (Everett 2012). Other observers similarly point to LRN's high-powered and connected board with being able to communicate with city leaders that genuine water-quality results could and should happen (Harper 2012b). Indeed, given the challenge of implementing TMDLs and reaching nonpoint sources, the Lynnhaven River TMDL is unlikely to have achieved as much as it has without the LRN–City of Virginia Beach partnership. In the same way, the partnership may not have acted as quickly and with such focus without the TMDL as a pressure-producing roadmap. The Lynnhaven River TMDL story is a striking example of how a local, nonprofit environmental group is able to work from a regulatory framework and support and assist local leaders in meeting it. Many TMDL-implementation plans remain sitting on the shelf; this one—thanks to LRN's insistence that meeting the plan was good for the river and watershed—did not.

The importance of the Lynnhaven River TMDL is also extending beyond its local watershed. In addition to serving as a national example at EPA, CBF has used it to support the larger, Bay-wide TMDL effort (Chesapeake Bay Foundation 2010). The Lynnhaven River TMDL also appears to have prompted ERP and CBF to partner together to work with the City of Norfolk to work on developing an implementation plan for a bacteria TMDL for the Lafayette River, a branch of the Elizabeth River that is primarily residential (Tribo and Katchmark 2012). Together, ERP and CBF convened approximately one hundred people from the Norfolk community to agree on an implementation plan to restore the Lafayette (Elizabeth River Project 2011a). As ERP has traditionally focused on commercial facilities instead of homeowners, the Lafayette project represents a widening of organizational focus (Tribo and Katchmark 2012). Perhaps, as Skip Stiles, the director of Norfolk-based Wetlands Watch observes, the ERP–CBF–City of Norfolk partnership is a sign of future collaboration among the city, local, and regional groups to solve some of the area's most pressing environmental problems (Stiles 2012).

FINAL THOUGHTS

Local watershed restoration does not happen in a vacuum. While many of the leaders and volunteers working on the local level are single-minded in their dedication to a specific river, the broader context also matters: from project funding to technical expertise, ERP, LRN, and NRPA are unquestionably benefited and informed by the larger agency and nonprofit context in which they work. Hampton Roads is an exceptionally rich region from which local watershed collaboratives draw expertise and human resources. Both VIMS at William & Mary and Old Dominion University's Department of Ocean, Earth, and Atmospheric Sciences have long provided critical scientific research support to both Bay and local restoration efforts. The fact that agencies such as NOAA and the Corps have a significant physical presence in Hampton Roads not only provides each of the organizations we studied tremendous opportunities for collaborative work, but also creates a source of retired personnel uniquely educated and experienced to contribute in a sophisticated manner to environmental restoration projects.

Our study also reveals that a diversity of nongovernment actors using different tactics benefits environmental restoration efforts more broadly. Having an organization with a more "adversarial" reputation such as CBF active in the area allows the local groups we studied to emphasize collaboration, create an identity in contrast to CBF, and establish trust with industry and government often suspicious of any group with an environmental agenda. (We understand that, from a regional and national perceptive, CBF is perceived as fairly moderate in comparison to groups such as Natural Resources Defense Council and Earthjustice; our point is that the local perception of CBF's "adversarial approach" allows the groups we studied to position themselves in contrast.) Indeed, organizational size may be a critical factor in determining whether an adversarial approach is the most suitable one. CBF's large membership, impressive financial resources, and regional presence allow it to press both industry and policy makers on a state and local level without risking key organizational ties, funding, or community clout. Local collaborations, on the other hand, excel at building trust among various stakeholders by using community ties, personal relationships, and a shared desire to improve the community's "backyard" (BIMBY). These very strengths make it much harder, however, for local collaborations to take a more aggressive approach, as such an approach could easily alienate key stakeholders and large segments of the community. Because the commitment, resources, and actions required to restore and preserve any river or watershed are challenging, our view is that diverse approaches—carrots *and* sticks—are needed and complement each other. CBF's more adversarial reputation nicely balances, for example, ERP's determinately collaborative one: both organizations achieve gains with their approach and benefit from the other's work and presence. As Stiles, executive director of Wetlands Watch observes, "[y]ou have to respect their organizational identity and the things that they do and don't do" (Stiles 2012).

Similarly, local watershed collaboratives need federal and state policy makers to provide a strong regulatory presence and framework so that they can work within and beyond the foundations set by regulatory requirements. As Wayland, a former EPA official, explained, "One thing we heard loud and clear from a lot of local groups was we need strong regulations. If we don't have strong regulations, there isn't a good foundation for us to build on" (Wayland 2012). Certainly, the Lynnhaven River TMDL is a striking example of how a watershed group used a regulatory framework to partner with stakeholders and make tremendous environmental gains. ERP experienced an arguably similar situation that led to the creation of its living trust fund. When EPA and the US Fish and Wildlife Service opposed Maersk's proposal to build a large port on the Elizabeth River, their aggressive stance allowed for ERP to play a facilitating role to create compromise. Without federal opposition to the terminal, ERP's reputation as a trusted collaborator may never have led parties to seek its facilitating leadership in the first place.

Finally, federal and state policy makers need local collaborations as well if they hope to accomplish many of the regulatory goals for stormwater and the Bay TMDL, as many of the actions that will be necessary to improve water quality require individual behavior change instead of traditional institutional controls. From Stiles at Wetlands Watch to Davis-Martin at DCR to Corbin at EPA, it is evident to all involved with the Bay TMDL and its implementation that government will not succeed if it decides to go it alone. Local watershed collaborations in Hampton Roads have already played a critical role in educating its members and the public about many of the BMPs that will be necessary to control runoff. The groups have also lent credibility to the Bay TMDL process generally, convincing local government officials that a genuine, grassroots desire for improved water quality supported the Bay TMDL effort and could help them meet the Bay TMDL's requirements. Put simply, local watershed collaboratives play a critical role in both legitimizing government action and engaging citizens to change their behavior. This role could not be more important for a process as complicated, ambitious, and fraught with controversy as the Bay TMDL. Certainly, it appears that some federal and state players are recognizing the power and importance of local watershed action. It will remain to be seen if they fully appreciate and enable its potential.

NOTES

1. At the time of the founding of ERP, Jackson was unmarried; her maiden name was Mayfield. To avoid confusion, we refer to her in this book as "Mayfield Jackson" in the text; interview material from her is referenced as "Jackson."

2. CBF filed a lawsuit in 2008 claiming Section 117(g) of the Clean Water Act *required* EPA to ensure that the Bay jurisdictions were meeting their agreed-upon reduction goals, but the case was settled after the Obama EPA agreed to implement the Bay TMDL (Houck 2011). The argument that the Section 117(g) created an enforceable instead of voluntary program was, as Oliver Houck, a leading expert on the Clean Water Act, put it "a bit of a reach" (Houck 2011). Many considered the case unlikely to succeed. Unless a new case is brought and a court determines Section 117(g) requires EPA to ensure Bay jurisdictions meet their promises as partners in the CBP, the CBP may fairly be described as a voluntary environmental program.

3. This arrangement allowed people to donate money to the NRPA and still realize a charitable deduction for tax purposes. CBF held funds destined for NRPA in a separate account, and did not charge NRPA a fee for providing this service. This practice was made unnecessary when NRPA was granted 501(c)(3) status.

4. The Bay TMDL cites as its impetus the requirements of two Consent Decrees settling the following lawsuits: *American Canoe Association, Inc. and the American Littoral Society v. EPA*, Civil No. 98-979-A (E.D. Va) and *Kingman Park Civic Association, et al. v. U.S. Environmental Protection Agency*, et al., No. 1:98CV00758 (D.D.C.).

FOUR

Restoring Princess Elizabeth's River

> *If you had asked me thirty years ago could the Elizabeth River be fixed . . . I would have said the last thing that would get fixed would be the Elizabeth River. So here is our great success story.*
>
> —James Oliver (2012)

An excavator operator brought his grandson to Money Point on the Elizabeth River in Chesapeake, Virginia, in 2011. They walked across a large field, an upland no-mow zone, toward the river. The grandfather swept his arm across the site, a project that included sediment removal, tidal wetlands, a riparian buffer, and a new oyster reef, turned to his grandson and said, "I brought you here because I wanted you to know what your grandfather does for a living. This is what I do; I restore the natural environment" (Stedfast 2012). A grandfather who operated an excavator on construction projects, proud of his work, wanted to be remembered by his grandson as a restorer of the natural environment. His grandson will remember that his grandfather participated in cleaning up one of the most polluted sites in the Chesapeake Bay watershed, if not the entire nation (Elizabeth River Project 2006).

Money Point was a thirty-five-acre ecological "dead zone" resulting from decades of industrial activity, particularly wood-treatment plants that used creosote as a preservative for bulk-heading materials, telephone poles, and railroad ties. Those plants dumped thousands of gallons of creosote at Money Point. Much of this came from a 1963 fire at the Eppinger-Russell wood-treatment plant. As a result the sediment at the bottom was loaded with polycyclic aromatic hydrocarbons (PAHs), a highly toxic organic compound that settles in the sediment and destroys much of the benthic population that is the foundation for the aquatic food chain.

It was not clear a few years before the Money Point project that such toxic legacy pollutant sites could be safely restored, nor was it understood how long it would take for normal life to return to the restored habitats if they could be restored.[1] Within nine months of completing the first stage of the Money Point project, a remarkable transformation had occurred. More than seventeen species of healthy fish were counted at Money Point by biologists with the National Oceanic and Atmospheric Administration (US Fish and Wildlife Service 2012). The years of work by the Elizabeth River Project (ERP) to make the river a better place had paid off with real and observable improvements.

Chapter 4 uses the model developed in chapter 2 (see figure 2.1) as a guide for examining how the ERP uses collaboration to develop and achieve its goals.

THE CONTEXT AND HISTORY OF THE ELIZABETH RIVER PROJECT

This section will discuss the political context and physical setting of the watershed and the Elizabeth River's history, followed by a discussion of the ERP's formation and first five years of work to develop a Watershed Action Plan to restore the river.

Political Culture

The political culture along the Elizabeth River was long dominated by traditional elites, the families whose aristocratic ancestors arrived in Virginia from 1630 until 1670 (Dabney 1971). Daniel Elazar describes this culture as paternalistic, elitist, and hierarchically ordered with the expectation that those at the top of the social order will take a dominant role in government, the primary objective being the maintenance of the dominant social order (Elazar 1984). That has changed in the twentieth century as the region became home to the greatest number of military bases in the world. With that change came a transient population that arrived from all regions of the United States, by and large stayed a few years, then left to be replaced with others arriving for a brief stay. The constellation of industries that developed to service the military and maritime industries, including rail, trucking, manufacturing, processing, and storage, influenced the transient nature of the population. The political order reflected this change by moving toward what Elazar describes as an individualistic political culture, one in which there is a higher emphasis on private concerns, where engaging in politics is an action to improve one's chances politically, economically, or socially. Social transactions "are based on a system of mutual obligation rooted in personal relationships" (Elazar 1984, 115). Mutual obligations can be established between individuals or harnessed through public, private, or nonprofit organizations to pursue and coordinate individual enterprise. It is pragmatic and programmatic, dismissive of ideological concerns while searching for solutions in the political marketplace. The outputs of politics are essentially a set of negotiated outcomes reflecting the diverse nature of the stakeholders.

The point here is that the cities in the Elizabeth River watershed appear to be a mixture of individualistic political culture grafted onto a traditionalistic political culture. The lingering effects of the traditionalistic culture may be a contributing factor to the somewhat tepid involvement of local government leaders in the watershed with ERP in its formative years (Adams 2012a). The participation that occurred during the early years, for example, appears to have been driven by personal involvement of individuals in city departments, rather than guided by the cities' political leadership. This is consistent with McNamara's (2010) point that personal relationships are most important in a traditionalistic culture. This has changed in recent years as Norfolk has partnered with ERP and the Chesapeake Bay Foundation (CBF) in cleaning up the Lafayette River, Portsmouth has taken on a large responsibility for the maintenance of Paradise Creek Nature Park, and Chesapeake partnered with ERP on the Money Point restoration.

The twenty-year effort of ERP to solve the problems of the river can be understood with this context of the individualist traditionalistic culture in mind, as it seeks to build consensus solutions with diverse stakeholders, one project at a time, in an environment that tends to be resistant to change. The diverse stakeholders each approach a new process with a particular set of interests. The outcomes are the result of diverse interests participating in the political marketplace, seeking a negotiated outcome that promotes both self-interest and the common interests of all participants. To be successful in this context, the environmental collaborative must be able to assume the role of facilitator-broker by convincing all interested stakeholders that it is in their interest, or at least not contrary to their respective interests, to participate in the process. In the exclusively traditionalistic political culture of nineteenth-century Virginia there would have been no process; traditional elites would have not allowed it. Today on the Elizabeth River, ERP is the fulcrum, the convener, of the search for consensus on cleaning the river.

The Physical Setting

The Elizabeth River, a tidal estuary, flows north through Chesapeake from its origins close to the North Carolina border at the locks in Great Bridge, forms the boundary between the cities of Norfolk and Portsmouth, and meets the James River and Chesapeake Bay approximately twenty miles to the north (see figure 4.1). The watershed, encompassing roughly 250 square miles, lies within the boundaries of four Hampton Roads cities: Norfolk, Portsmouth, Chesapeake, and Virginia Beach. The main stem flows north from downtown Norfolk and Portsmouth for almost seven miles to meet the James River. Three tidal branches feed into the main stem. The southern branch flows north from Chesapeake, the eastern branch from Virginia Beach, the western branch from Portsmouth. There are no free-flowing tributaries into the river. Two miles wide at its mouth, the river narrows into a channel that is wide enough to accommodate pleasure craft and barge traffic on the intracoastal waterway at the Great Bridge locks twenty miles to the south.

Ship repair yards and industry line the eastern branch for two and a half miles. Beyond that point the river is residential, where it accommodates small craft for another three miles. Beyond, the river becomes a shallow stretch of salt water marsh for another two miles. It stops short of the colonial-era port of Kempes Landing, which used to accommodate ships at its tobacco warehouses at what is today the intersection of Princess Anne, Kempsville, and Witchduck roads in Virginia Beach (Frankenberry 2011). The headwaters are at a dam that separates it from a barrow pit dug to build a freeway interchange on the Virginia Beach-Norfolk Expressway at Witchduck Road. The western branch reaches seven miles into Portsmouth and is residential. The Lafayette River branches east from the main stem for four miles into the northern part of Norfolk, with residential neighborhoods along each shore.

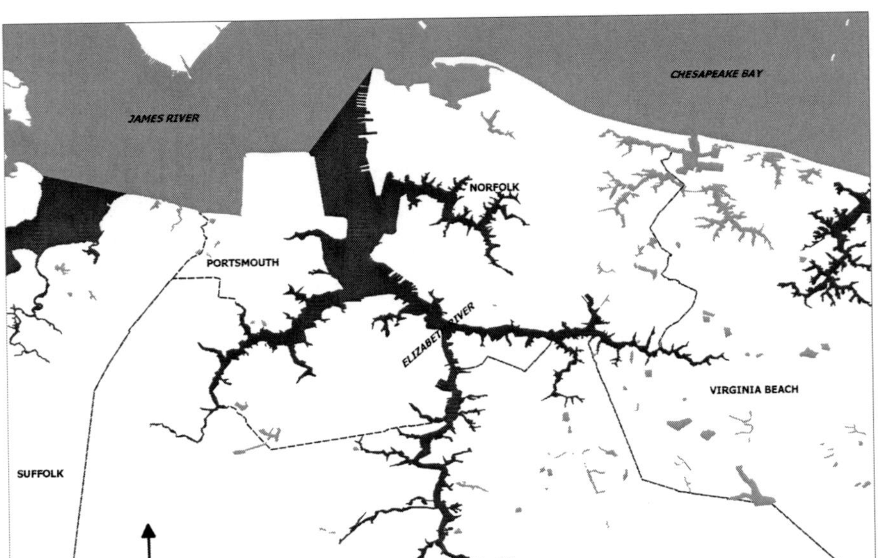

Figure 4.1. The Elizabeth River Watershed

The Colonial Era River

Captain John Smith and a crew of twelve entered Hampton Roads and the Elizabeth River in 1608, the year after they founded Jamestown, several miles to the northwest on the James River (Yarsinske 2007). They went several miles south on the river and made note of the forests of fir and pine along the banks. Eleven years later, in 1619, the river was named after Princess Elizabeth, daughter of King James I. The following year, 1620, a John Wood apparently recognized the value of the nearby pine and fir forests and petitioned the London Company to build a shipyard on the western shore of the southern branch. That same year another resident petitioned for land on what is today the world's largest naval base, Norfolk Naval Station. Land was purchased and the site laid out for Norfolk in 1680. Land was purchased for Portsmouth at the same time, though the site was not laid out until 1716. As the Revolutionary War approached, land was purchased by a Portsmouth merchant along the western bank of the southern branch to build a shipyard. The Gosport Naval Yard, built in 1767 to service both commercial and British ships, is today Norfolk Naval Shipyard, the oldest ship repair facility in the nation (Yarsinske 2007).

During the 1700s, Norfolk, established along one of the finest natural harbors in the world, became the leading center for trade for the Chesapeake Bay region, sending tobacco and other agricultural commodities to England in exchange for the products manufactured there. The river also became a major center for ship building. The scene on the river was described in the mid-1700s as "gazing out on the Elizabeth . . . one saw a river choked with merchant vessels"(Yarsinske 2007). The Elizabeth offered more than a port facility bustling with commerce, ship building, and repair industries. As with the Chesapeake Bay and its many other tributaries, it provided a bountiful harvest of fish and shellfish:

> A writer in 1679 described the Elizabeth River as pristine and filled with many species of fish. And it was noted that oyster reefs in the river were so large that at low tide they were a foot above the water and easily harvested. (Maddry 1996, E1)

The Modern River

From such auspicious beginnings, the river, one of the largest natural harbors in the world, developed to become one of the biggest and busiest industrial harbors and commercial ports in the world. It is a world-class harbor as well as a historic waterway. It is also an urban, industrial river befouled with pollution and it remains a struggling river in the twenty-first century. Traveling south on the river from its mouth at the James River takes one past military bases, commercial container terminals, coal facilities, ship repair yards, and other industries that line the shore and are the core of the region's economy. Begin at the mouth of the river by the world's largest naval base, Norfolk Naval Station. Travel south for three and a half miles past the Naval Station and one arrives at Norfolk International Terminals, the Port of Virginia's largest bulk cargo terminal on the river and centerpiece of one of the world's busiest container ports, lined with fourteen impressive cranes that are among the largest and most efficient in the world. Just to the south, past the mouth of the Lafayette River, are the Lambert's Point Coal Terminals, the largest coal-exporting facility in North America. Across the river from the coal terminals are the APM Terminals, the largest privately owned terminal in North America (APM Terminals 2012). The Port of Hampton Roads ranks third among East Coast ports in container cargo volume with the combined annual activity of APM Terminals and Norfolk International Terminals (College of Business and Public Administration 2010).

Beyond this point, where the river separates downtown Norfolk and Portsmouth, begins the southern branch. Here one passes by several shipyards and other high-density industrial developments that line both sides of the river. There is no public access here where wetlands have been filled and most of the river is bulkheaded. It has been dredged to double its normal depth and is only two-thirds the width it was when John Smith sailed into the southern branch in 1608. The deepening and narrowing of the river for industrial use reduces the river's tidal flushing and further burdens the ecosystem. The fir and pine forests that encouraged John Wood to purchase land to build a shipyard on the Portsmouth side of the river in 1620 disappeared long ago, further eliminating barriers to stormwater runoff. This is an industrial river and, because of the particular types of industry that made this the center of economic activity in the state so long ago, it has suffered severe ecological decline. Perhaps most important has been its use as a repository for several of the most toxic chemicals known. That legacy has given the Elizabeth River watershed the notorious reputation of having seventeen sites on the US Environmental Protection Agency's (EPA) Superfund list. It is so impaired that the EPA has a remediation team established just to monitor cleanup efforts on the river (Virginia Department of Environmental Quality 2012a; USEPA 2012c). Yet there was little public awareness of this great, historic river or its problems among the citizens who lived in Hampton Roads at the beginning of the 1990s. Robert Bray, former executive director of the Virginia Port Authority, who grew up on the river, said, "I don't think it was even on the radar—here, boats came and went and that was the end of it" (Bray 2012).

The Polluted River

The Elizabeth's southern and eastern branches are severely impaired, the main stem and Lafayette are impaired, and the western branch is marginal (Healthy Harbor Case Studies 2010).

The industrial development along the Elizabeth, the engine for economic development in the region beginning four hundred years ago, made it one of the most polluted rivers in the Chesapeake Bay watershed and the entire United States. The industrial activities along the river, particularly ship repair and the military, are by the nature of their activity polluting and for much of their history they used the river as a dump. Whereas the industries on the river today are heavily regulated by the EPA and Virginia Department of Environmental Quality (DEQ) for point source pollution, the river contains several badly damaged sites from legacy pollutants, deposited before the modern era of environmental regulation. Those legacy pollutants are the source of most distress and damage to the ecosystem, particularly the benthic communities at the bottom that are the base of the food chain (Virginia Department of Environmental Quality 2012b). They include toxic deposits from several wood-treatment plants on the southern branch of the river as well as the toxic residue from the many ship repair yards.

Among the several ship repair yards that operate on the southern and eastern branches, Norfolk Naval Shipyard in Portsmouth, founded in 1767, is the country's oldest and largest ship repair yard, as well as the US Navy's largest industrial facility (Yarsinske 2007; Norfolk Naval Shipyard 2012). It operated as a British naval yard until the Revolutionary War, when it was confiscated by Virginia, only to be burned by British troops in 1779. The federal government bought the shipyard from Virginia for $12,000 in 1801. The Confederacy took control of the shipyard during the Civil War and rebuilt the USS *Merrimack* there. The Confederates burned the shipyard when they retreated in 1862. It underwent a major expansion during World War I and World War II, when it employed as many as forty-three thousand personnel and constructed 6,850 vessels (Norfolk Naval Shipyard 2012). During the same time it expanded south from its original site to Paradise Creek, which snakes west into Portsmouth for 2.6 miles, with industry along the north side and the neighborhood of Craddock on the south side. During the two wars the most convenient way to dispose of waste from its repair work was to remove it across the short distance to dump it along Paradise Creek. Three Superfund sites were designated along Paradise Creek as a result (Elizabeth River Project 2008b; Chesapeake Bay Foundation 2006; Reisch 2012), sites since cleaned up by the Navy, which during the cleanup went beyond the regulatory requirements by digging up and removing the contaminated sediments instead of simply capping them and at one site restored upland wetlands, replacing nonnative plants with local plants a good distance beyond what was required (Elizabeth River Project 2008b).

The waste materials from the naval ship repair yard and several other ship repair yards include abrasive sandblast material; highly toxic tributyltin, an antifouling paint additive once used on most naval and commercial vessels and now banned by international agreement (Unger 2012; Ewing 2012a); calcium hydroxide, a byproduct from producing acetylene, polychlorinated biphenyls (PCBs), PAHs, mercury, and other assorted heavy metals and organic compounds.

Residential neighborhoods and agriculture add to the pollution load. Excessive nutrient runoff from residential neighborhoods across the watershed and from agriculture in southern Chesapeake adds to the river's pollution diet. Large algae blooms appear in warm-weather months, depleting the dissolved oxygen necessary for marine life to thrive. This is a particularly serious problem in the Lafayette tributary. The Lafayette's main problem is the nutrient load from stormwater runoff and high bacteria levels from sewage and waste from pets and the migratory waterfowl that have increasingly made the region a permanent home.

Formation of the Elizabeth River Project

The Elizabeth River was perceived by many to be a dead river in 1991. It appeared from conversations with Hampton Roads Planning District Commission leaders that their intention was to lobby to get the river declared an industrial river, which meant control over the river would be lost (Adams 2012a; Moses 2012). It would be an assertion that the river was beyond saving and therefore beyond need of regulation (Adams 2012a; 2012b). The river would simply become a dump for the pollution output of all users.

Four concerned citizens met shortly thereafter at the home of Sharon Adams in Virginia Beach, in November 1991. As they sat around a kitchen table discussing the river and their role as environmental advocates, they agreed that something should be done to give the river a chance before consigning it to industrial river status. Marjorie Mayfield was a reporter for the *Virginian Pilot* who had taken a sabbatical for several months. She lived on Scott's Creek along the Elizabeth River and spent a lot of time in her backyard close to the river during those months. Whereas the public and the Planning District Commission appeared to have written the river off for dead, she watched herons feeding at the beautiful little stretch of river from her backyard that was filled with life. Reflecting on her work and life, she decided that what would really matter to her would be to clean up the river that was in her backyard (Jackson 2012a). She quit her job as a reporter and took a job as a waitress and freelanced as a writer.

Adams was executive director of the Virginia Beach SPCA. She had been deeply committed to environmental advocacy for many years, having served as the president of the Norfolk-Virginia Beach League of Women Voters (LVW) and as the League's Virginia environmental advocate for several years, as a founding board member of Southeastern Association for Virginia's Environment (SAVE), and as a member on Governor Gerald Baliles's Chesapeake Bay Land Use Round Table that spent two years developing the legislative proposal for what became the Chesapeake Bay Preservation Act. She had represented the LWV on roundtables facilitated by the University of Virginia's Institute for Environmental Negotiation that developed legislative proposals on a phosphate ban in laundry detergents and a low-level nuclear-waste policy, both enacted into law by the Virginia legislature. Robert Dean had received several awards and national recognition for his environmental activism. He founded Clean the Bay Day in 1988 and for several years recruited, organized, and managed several thousand volunteers to clean up debris along the shorelines of the Bay and its tributaries in Hampton Roads one day each year. He had worked with Adams in SAVE. Mayfield knew Robert Dean. Adams and Dean talked about the river. Mayfield and Dean talked about the river. They decided to meet and discuss options. Michael Kensler had recently moved to Hampton Roads to open the first CBF office in the region and was invited to the meeting because of the Foundation's stature on the Bay.

They made several decisions that night. First, they understood that powerful and diverse stakeholders representing businesses and governments depended on the river. They had to identify them and talk to them about the river to find out how they felt about it. They also needed to know if anyone among public and private stakeholders thought the river was worth trying to save.

> I felt this enormous burden that somehow this was the last-ditch effort and if we messed up it was done. Therefore we had to try to get it right. We had to under promise and over deliver. We had to be very careful what we said we were going to do. Four people with no money and no power. The idea was to build up respect for the river. Practically speaking it didn't make any sense and economically it didn't make any sense. There was no argument that could be made other than the argument about your internal feelings. There was no practical reason, except pride and affection and that it was just morally the correct thing to do. The minute you tried to play the game of blaming folks about how we got into this mess, you were going to drive people into a corner and we knew that because that is what environmentalists do to some degree. (Adams 2012b)

Second, they agreed on what would become the ERP's mantra during the ensuing decades. There would be "no finger pointing":

> One of the key decisions that first day was that if we are going to do this with the river it is going to be collaborative. There are so many powers on the river. It contains the world's largest navy base, world's largest coal exporting facility, ship repair companies. We cannot point fingers. It has been something we have never really wavered from. (Jackson 2012a)

Third, they decided they needed money for the interviewing and to prepare a study documenting regulations affecting the river. They received small grants from Clean the Bay Day and the Virginia Environmental Endowment to conduct the interviews and the regulatory study. Adams knew Jerry McCarthy, executive director for the Virginia Environmental Endowment, which funds environmental projects and research in the state. She called him and talked about their ideas and possible funding. They agreed to meet and the four drove to talk with him in Richmond in January 1992; they were awarded a mini-grant (Elizabeth River Project 1992):

> I guess Sharon called me up—and said, "I got a great idea for you. We're forming a group and we need your help to get it off the ground." So, they came—the four of them—to see me in Richmond, and we sat around the table in my office and had quite a discussion, and I was so tickled—these people who claimed that they'd dreamed this idea up while drinking beer and eating pizza a recent night beforehand, waking up the next morning realizing they'd actually agreed to restore the Elizabeth River and if they were serious about it, they better come up with some money and a plan—so they came to see me and spent an hour explaining how and what they wanted to do and I was really intrigued—and saying oh man, I wonder how much money they want—so finally I asked how much—and Sharon says $1,375. And I said DONE! Just like that. Just get me some paperwork. So we made that grant fast before they changed their minds. (McCarthy 2012)[2]

The four spent the next year and a half talking with people and building a consensus for doing something to restore the river, and gathering information for the purposes of developing a long-range plan for restoring the river and increasing public awareness of its economic, cultural, aesthetic, and recreational value. Sixty-five individuals representing decision makers, most of whom had an economic interest in the river, including public officials, regulators, business and industry, the military, and scientists were interviewed by the founders between March and August 1992. They were joined in the interview process by Katherine Cross, a Norfolk attorney, and Mary Heinricht, an environmental planner (Elizabeth River Project 1992). They talked with the interviewees about the historical, economic, and cultural significance of the river. They wanted to know the following: (1) Do people care; (2) Can the river be fixed; (3) Would they participate (Adams 2012a; Elizabeth River Project 1992):

> We went into meetings assuming that we could come out having made a friend. We used the argument of history, of culture, of pride, that the participants in the process could be keepers of the tradition. (Adams 2012a)

Most of those interviewed spoke of the economic value of the river, the river as the lifeblood of the local economic system; but many wanted recreational activity and a healthy river to coexist along with the economic activity on the river. There was a general sense among those interviewed that the public was not aware of the river as anything other than a dumping ground. Most important, those interviewed agreed that it was worth the effort to try to save the river and were willing to participate. The respondents were interested in pursuing a cooperative, consensus-based approach, and educating the public about the river was considered important by many. Since other studies and agency-led efforts to help the river had been ineffective, it was understood that this project would need involved stakeholders from all sectors to have a chance for success; not just their trust in the process, but real enthusiasm for the process of agenda setting, action planning, and implementation. Two broad goals were developed from the interviews. The first was to "establish [a] cooperative, inclusive project." The second was to "foster public education and awareness" (Elizabeth River Project 1992). Though the second goal would appear to be of the upmost importance for restoring and maintaining the river, it would be several more years before educating the public received much attention from the ERP board and staff, though every board member and collaborating partner interviewed for this project gave it the highest priority.

An advisory board of thirty, representing the diverse range of stakeholders, was selected from among those interviewed who were enthusiastic about participating, offered knowledge and creative visions of what was possible, and were connected to a particular constituency (Elizabeth River Project 1992). All stakeholder groups were represented on the advisory board, which held its first meeting in September 1992. The agenda was twofold: (1) discuss ways to educate the public about the river, and (2) discuss ways in which consensus could be used. The governing board, comprised of ten members including the founders, was surprised when the advisory board urged a more aggressive schedule than the project team had in mind. The advisory board pushed for a second meeting sooner than the six months proposed by the governing board and they urged faster movement on developing approaches to improve the river's ecological health. It was in this first advisory board meeting that someone suggested the goal of "children swimming in the Elizabeth River by 2020" (Elizabeth River Project 1992). That seemingly elusive goal gradually became a real option and was embraced in the revised Watershed Action Plan in 2008 (Elizabeth River Project 2008c). The enthusiasm developed at the meeting gave the project needed momentum to push to the next phase and affirmed a real desire on the part of river users to improve its quality. The pace accelerated faster than envisioned by the founders, and that was positive.

The governing board/project team spent most of 1993 building consensus and developing an agenda. The ERP incorporated as a 501(c)(3) nonprofit organization in early 1993 with a mission "to restore the Elizabeth River to the highest practical level of environmental quality through government, business, and community partnerships" (USEPA 2008). Marjorie Mayfield Jackson was appointed executive director on a part-time basis. Adams was elected to be the first president. The ten-person governing board included the four founders. One of those governing board members played a critical role for the ERP in 1993 (Jackson 2012b). Dr. Robert Huggett of the Virginia Institute of Marine Science (VIMS), a renowned scholar who had done much of the seminal research on the effects of contaminants on the fish population in the Elizabeth River and who had discovered the cancers prevalent in the mummichog population, joined the governing board. He was a member of the EPA's science advisory board, had extensive contacts there, and was familiar with the EPA's new comparative risk-analysis approach. He recommended that ERP consider using it. Huggett and Mayfield Jackson went to Washington, D.C., and met with the head of the Comparative Risk Analysis Branch of EPA's Office of Policy, Planning and Evaluation. They received a small grant to explore comparative risk analysis, a process best described as the marriage of science and public values, as a method to establish ERP priorities (Jackson 2012b; Elizabeth River Project 1995).

Members met for much of the year in 1993 to develop the guidelines for the conference scheduled for October. The comparative risk process was launched at that first large conference of stakeholders at the Norfolk Waterside Marriott in downtown Norfolk on October 22, 1993, after almost two years of dedicated preparation by the founders, with substantial input from stakeholders. The conference, "Elizabeth River Visions," was attended by 225 people. The participants discussed and debated competing uses and values for the river, then broke into groups and talked and listened. By the end of the meeting about eighty participants had volunteered to sign on to work with three committees during 1994 to assess and rank the river's problems; a citizen/industry committee, a government/agency committee, and a science/technical committee (Elizabeth River Project 1995). Joseph Coccaro, writing in the *Virginian Pilot*, described it as follows:

> Do you leave it alone? Or do you risk making matters worse by trying to clean it up. A group of unlikely bedfellows concerned about the Elizabeth River has joined together to address these questions and form a consensus on what should and can be done to restore one of the

> nation's most abused waterways. The group calls itself the Elizabeth River Project. It consists of state environmentalists, water quality regulators, shipyard executives, scientists and military representatives—individuals who might otherwise be at odds over water pollution issues. "We do have a problem in the Elizabeth River," said Bob Huggett, a Virginia Institute of Marine Science professor and Hampton Roads' resident guru on the river. "One of the things that will hopefully come out of the Elizabeth River Project is some structure of desire and needs." (Coccaro 1993)

The meeting at the Marriott, two years after four people had a conversation around a kitchen table about what they could do to help restore the river, was a pivotal event for the ERP. Important public stakeholders from government, industry, the military, science, and nonprofits were represented. They worked efficiently and signed on to a serious commitment of personal time to continue the process for at least another year. Equally important was that participants sat and talked with people from other sectors, which for many had not occurred before. People listened carefully to what others had to say and they learned. Several interviewed for this study were at that meeting many years ago and think that it was a seminal event in terms of the early and fairly rapid development of trust and social capital (Bray 2012; Host 2012; McCarthy 2012; W. Priest 2012). The process afforded participants an opportunity to talk, listen, and learn. The two years of effort to prepare for the event had been well rewarded. Decisions were made, responsibilities assigned and accepted, scientific knowledge was made available for the next steps in the process, and the process itself had gained far more traction than anyone might have expected at the beginning of the day. Mayfield Jackson had said before this meeting:

> One of the first decisions we made was not to be adversarial, not to point fingers—to recognize that this is Virginia, and it doesn't work that way. For any real change to happen, it can't be the regulators saying this is what needs to happen, and it can't be the environmentalists saying that. It has to be the whole community. (Swift 1993)

The diverse stakeholders from the community were, by and large, now engaged and prepared to move forward. The ERP received $50,000 from the EPA and $20,000 from the Virginia Environmental Endowment to conduct the two-year comparative risk assessment and develop risk-management strategies to address the problems (Elizabeth River Project 1995). A description of the two-year effort follows to demonstrate the thoroughness of the risk assessment and risk-management processes undertaken by ERP.

Early in 1994 some eighty volunteers, divided into three committees, began the difficult and time-consuming work of investigating the risks. To use the comparative risk approach the groups were asked to consider the best science, on the one hand, and the competing values among stakeholders, on the other (Elizabeth River Project 1995). The committees convened for an overnight retreat in late March 1994 to hear scientific presentations on the river and begin building a common understanding of the data that would be important to an effective outcome. They held small group meetings with different stakeholder interests represented in each group.

The three committees met separately in April to further develop and refine their specific issues (Elizabeth River Project 1995). A month later, in May, the full group reconvened to begin the work of integrating the issues. They winnowed the issues by focusing on those where data existed for quantifying the risk. Hence two issues, the lack of public education and public awareness, were put aside for the risk-assessment cycle, but not dismissed; it was agreed that they could be taken back up in the risk-management cycle. A technical report committee was established to develop technical reports and provide a scientific and data-based assessment of the risks to the ecosystem, human health, and quality of life, for each identified risk. A consultant from the Oak Ridge Center for Risk Analysis helped develop guidelines to ensure the technical quality of the reports. First drafts of the eleven technical reports were completed for presentation at a late August fish fry in Norfolk attended by about one hundred ERP supporters. A peer-review process followed and final drafts were submitted for circulation at the end of September. Further refinement of the degree of risk for each issue took place over the next two weeks; included was consideration of the impacts on the economy and aesthetics.

Finally, in late October 1994, the three committees met for another full-day session and reached consensus on the criteria for the final ranking session. That session occurred on November 17–18 at the US Coast Guard Support Center in Portsmouth, fittingly on the Elizabeth River. Final agreement was reached on the ranking of risks at that meeting. The one issue on which consensus could not be reached was point source pollution. Some argued that it was under control because it was so highly regulated; others argued that whether regulated effectively or not what was being discharged legally into the river was still highly contaminated and a high risk. This issue was the one issue that had to be settled by voting rather than consensus. It was in the end ranked among the four highest risks, which were as follows:

- sediment quality and sedimentation processes,
- loss of habitat and biota,
- nonpoint source pollution, and
- point source pollution.

An issue that would compete for the ERP's attention along with these high-risk factors in the years to follow was the lack of public awareness about the river and its problems throughout the region and the need to educate about the river (Elizabeth River Project 1995).

The planning work remained incomplete, however. In January 1995, the ERP held its second annual conference and announced the decision on risks (Harper 1995). The rest of that conference was spent discussing what to do about the risks, moving the group for the first time beyond risk assessment to risk-management planning. A 120-member Watershed Action Team was established to draft a comprehensive plan for restoring the Elizabeth River. Four task forces were created to address the four highest priority issues, research ways to handle them, and report back in the spring of 1996. The four task forces were:

- Sediment Quality Task Force,
- Habitat and Living Resources Task Force,
- Water Quality Task Force, and
- Toxics Reduction Team.

The one remaining action in a five-year process of consensus building was to unveil the Watershed Action Plan at a coming-out ceremony, an event that needed to be highly visible to the public (Elizabeth River Project 1996). It took place at Nauticus in Norfolk on June 20–21, 1996 (Moses 1996).[3] Popular celebrity newscaster Charles Kuralt emceed the event set against the backdrop of the river. The ERP was going public, so to speak, with an eighteen-point plan to restore the Elizabeth River to the highest level while retaining its economic vitality. The five years of building trust and social capital through a consensus process, researching, planning, organizing, preparing was ending. The way forward would have to involve "doing," while continuing to build social capital.

> That was the launching pad for the organization. It got a lot of attention at that point. Funding became serious. It was a stressful time and there was a lot of tension around what we were doing and how we were doing it. I think the feeling was that we were serious now and we needed to be sure that everything we did from that point forward was credible, honest. We had taken on the river and had this obligation. (Adams 2012b)

Thus, the formative years for the ERP came to an end. It was a mature organization now. The process described here was a highly labor-intensive effort by a large group of talented and well-educated persons, who committed a substantial part of their lives to the comparative risk assessment process for two years to get to the publication of the ERP's plan for river restoration. A large network of competing stakeholders had become convinced that they could work with one another through the ERP, which would sustain the vision, broker the consensus agreements, and make substantial changes in the river. The ERP itself would change in some fundamental ways; it would grow quickly in full-time staff and annual budget and soon take on the business of habitat restoration and develop and manage the highly participatory, celebratory River Star programs, thereby involving the industrial players in the process. It would bide its time, working on small projects, continuing to build credibility, and building social capital to get to the very dramatic sediment-remediation projects that were at the top of the list of risks to the river, identified in the exhausting two-year process recently concluded. It had taken more than five years to get from the kitchen table to Nauticus. Though it was a record of substantial achievement and success, the first real project on the river had yet to be started.

COLLABORATION PROCESS

The four founders understood at their first meeting that the only chance for success would be to get all stakeholders to participate by making it nonthreatening. They would not talk about blame for the river's problems so as to avoid alienating potential partners. They paid particularly close attention to this principle as they talked to the stakeholders they identified as critical for inclusion in the project. That project was to bring stakeholders into a process to change the behaviors of those who used the river and thereby achieve a cleaner, healthier river, on the one hand, while creating greater public awareness of the river as an aesthetic, historical, cultural, and economic treasure, on the other. They would always frame discussions with involved parties as a search for a win-win solution.

The Elizabeth River presented a complex set of issues and stakeholders. They were aware that this would be a long-term effort that would necessarily develop slowly, if indeed it developed at all. That first big decision to forego the blame game and to implement a nonthreatening, consensus-based process, described earlier in this chapter, was followed so successfully during the formative years that it was embedded in the organization's culture by 1996, when the first Watershed Action Plan was presented to the public by Charles Kuralt.

After five years of building trust and social capital, the ERP began to shift its focus to three new efforts: project development and implementation; program development to more fully engage industries along the river; and extending the organization into the business of education to heighten public awareness. As it pursued these new directions, the search for win-win solutions was a principle that would not be compromised. Every person interviewed for this study, in explaining what made the ERP different and successful, responded that it was the collaborative approach, or as they all put it, the "no pointing fingers" approach, always searching for consensus:

> It is a slow consensus-building process. They are deliberate about that stuff. I remember there would be times in some of those meetings where you are like, "You guys are still debating the rough draft of the action plan." It is very methodical to get a consensus. It takes a lot of time. It's a lot of meetings. I have been to those meetings and it does take a lot of time and they have found people that are willing to make it happen. (Harper 2012a)

Retired Regional DEQ Director Frank Daniel explained why DEQ had enough faith in ERP to trust it to lead and manage a scientifically and technically complex project such as the Money Point clean up:

> Well, you have to go back to what the Elizabeth River Project is. It's a partnership, it's a collaborative effort. One of the many things that always attracted me to the Elizabeth River Project is the fact that there's no hidden point. So that, early on it was established in a number of the early meetings to help us put together the Elizabeth River Project that we didn't want to be into finger pointing. I mean, you can certainly find a lot of entities along the river that you could point fingers at. We wanted everybody to come into the partnership, or at least be invited to come into the partnership, without finger pointing, without saying, "You caused this, therefore we need you to do this." Everybody come together, bring the best of all of your minds and try to come up with a way to help speed up the recovery of the river. And to do that you really have to step away from finger pointing and so that was really important. If we are to manage the very technical aspects of a lot of these cleanup projects, particularly when you get into sediment remediation, you would have to have a collaborative effort. But people in ERP alone couldn't do that without all the help they had, collaboration from EPA and DEQ and Army Corp of Engineers and others, such as VIMS and ODU scientists. So they had a tremendous number of scientists and engineers bring their intellect and their minds into that partnership to help, but ERP's also been lucky in that they've been able to hire some very good people like Joe Rieger and Dave Koubsky, people who have scientific and engineering backgrounds. And they've been able to manage that process. So you've had a lot of people coming together in this partnership really without looking to get any praise themselves, without wanting to point fingers, just everybody really being dedicated to help bring the Elizabeth River back. (Daniel 2012)

ERP's approach provided a safe platform for participants. Predictability and dependability have been found to be essential for environmental collaboratives to achieve success (Wagner and Fernandez-Gimenez 2008). ERP established a rock-solid reputation for both; partners in the collaboration process knew their opinions would always be listened to with respect. The approach attracted stakeholders and was the glue that kept them involved and sustained the ERP's efforts. And as already discussed, the opportunity to meet in 1993 at the Marriott, with stakeholders who brought quite distinct and different interests into the meeting, was a significant event for participants and the ERP. It was the first chance for many who attended to meet and engage in a legitimate sharing of values and interests with those who they otherwise may not have met, or perhaps had met only in adversarial circumstances. The process of collaboration itself appeared to become a driving force for further participation with the ERP for many who attended that meeting (Bray 2012; Host 2012; McCarthy 2012; W. Priest 2012). The culture of collaboration would develop and deepen in the second generation of ERP efforts with such programs as River Star Industries and River Star Schools, and set the stage for high-cost, high-profile projects such as the Money Point restoration and the Paradise Creek Nature Park project in the ERP's second decade:

> We have used the same model for all of our focus areas. Our purpose was for people to listen and for the plan to be theirs. We pay professional facilitators so that our absolute goal is to hear from everybody and to empower them because we think that is how plans get implemented. (Jackson 2012b)

The use of facilitators began with the first large meeting in 1993 at the Waterside Marriott. It has been used for every planning process since, most recently to develop the Lafayette watershed plan (Jackson 2012b).

Stakeholders and Roles

Several groups of stakeholders were identified as essential to a collaborative effort to help the river. Industry was the first. Even though unregulated nonpoint source pollution from stormwater and agricultural runoff may be the most complex problem to solve on the river, the maritime industry had left the major legacy sites polluted while it provided for the region's economic vitality. Although the founders learned that everyone they initially interviewed was willing to continue the discussion about participating in decisions to make the river cleaner, the main stakeholder, industry, joined with some trepidation:

> Business people in the beginning and for a long time participated somewhat reluctantly. They didn't know what was going to come out [of the process] and they wanted to be part of the decision. People would come in that way and then they would get caught up in the excitement of it and had the good feeling of making a difference and that still happens. (Jackson 2012a)

The US Navy was another important stakeholder. Its actions tended to have similar polluting effects as the ship repair yards. It had the largest naval base in the world along three and a half miles of shoreline. On the southern branch it had the largest public ship repair facility in North America. The activities of both had resulted in several Superfund designations. The Navy was a groundbreaker for starting and successfully completing several Superfund and potential Superfund cleanups (W. Priest 2012; Reisch 2012). It became the first Model River Star (Barnett 2012).

The Elizabeth River is a river of industrial, maritime, and military users; it is therefore heavily regulated by federal and state governments. The watershed is in all or parts of four major cities. These conditions drove the selection of stakeholders to participate in the ERP. The industries and federal government owned much of the land along the river and had resources needed to help fund restoration projects. Thus industry, state and federal governments, and regulators were identified as priorities for those first interviews conducted to measure the level of interest and willingness to participate. The project would succeed or fail on their willingness to join the process and remain engaged. Local governments were also included, though private industry and the federal government owned most of the land on the river. Local governments did not have the authority to write ordinances affecting the river; federal and state governments wrote the regulations. But local government endorsement and expertise were very important for the project.

The scientific community was also crucial to the project's success. The river's problems were complex from a scientific point of view and the founders were not scientists. Scientists would serve at least three functions. First, they provided the science that would shape part of the discussion among stakeholders. Second, they were the project experts for implementation, making sure that any effort undertaken by the ERP was built on the correct science. Third, if they were successful on the first two, they would bring much needed credibility to ERP in the broader community, a function they began to fulfill quickly by helping bring the EPA into the process (Jackson 2012). Civic leaders who had a particular interest in the river were also interviewed and some of those continued to participate. Among the sixty-five individuals interviewed in the first part of 1992, the two groups with the most representatives were local government officials with fifteen and the maritime industry along the river with fifteen. Other groups were state officials, particularly the regulatory community (eight), civic leaders (seven), nonmaritime industries along the river (four), and three scientists from local institutions of higher education (Elizabeth River Project 1992). Mary Heinricht (2012) said of industry participation:

> I think it was clear that industry was being given the opportunity to lead the process and not be a victim of it. I am not sure it had ever happened before. I think they started out at the beginning of the meetings and nobody trusted anybody.

The number of scientists increased to eight in attendance at the 1993 meeting at the Norfolk Waterside Marriott that launched the comparative risk process (Elizabeth River Project 1993). The three comparative risk assessment teams that finalized the risk-assessment rankings included thirteen scientists from VIMS, Old Dominion University (ODU), and Tidewater Community College (TCC) as well as others trained in science and engineering who worked with government (Elizabeth River Project 1995).

The ERP's mission is: *To restore the Elizabeth River to the highest practical level of environmental quality through government, business, and community partnerships.* Notable for their absence in the mission were citizens and residential property owners, unlike the other cases in this study. The problems on the Elizabeth result from industry, which owns, along with government, most of the land; the exception is the Lafayette tributary in Norfolk, which is primarily a residential river. The ERP is thus unique among the groups in this study, since in the Lynnhaven and Nansemond watersheds the primary problems result from the activity of homeowners and agriculture. The nature and magnitude of the problems on the Elizabeth required a collaborative approach focused on the industrial uses of the river with industry as the core stakeholder group. That focus necessarily left out some elements of the Hampton Roads community, at least for the first two decades.

It would be twenty years before the ERP, partnering with the CBF and funded with a $103,000 grant from the Hampton Roads Sanitation District, set out to get citizens along the Lafayette River to change behaviors that pollute the river (Elizabeth River Project 2011b; Harper 2011a; Harper 2011b; Clancy 2011). The River Star Homes program was launched in 2011, two decades after the original meeting around a kitchen table, to help residents and homeowners in the Lafayette watershed implement best management practices for lawn care and controlling pet wastes, identified as two major sources of nitrogen and bacteria (Harper 2012a). Mayfield Jackson (2012a) explains:

> In the early years we did not really focus on the citizens because there were too many of them. To try to reach five hundred thousand people one by one would take too long. We focused on the people who could make the biggest difference, which was industry and government. Lately, in just the last year, we have turned to finally look at the citizens with this River Star Homes program, which now has seven hundred homes.

Conveners

The four founders were the first conveners. None were local business or government leaders who could call up other business and government leaders and be instantly credible. Adams characterized it as "four people with no money and no power" (Adams 2012). What they did bring to the table, literally meeting around a kitchen table for the first meeting, was a desire to make a difference for the Elizabeth River. They brought a vision, great energy and commitment, as well as a mix of appropriate skill sets.

Mike Host, who served on the ERP board until recently and was then the environmental affairs officer at Norfolk Naval Shipyard, participated at the Waterside Marriott in 1993 in the first big meeting that was the beginning of the comparative risk assessment process. He had this to say about the founders:

> In that first meeting—wow, what a divergent group of individuals. I said no way—and everybody just listened—no matter how different our thoughts were. It really ended up being a product of our collective thoughts and we were all proud of it. Very few things in my life have felt quite that special in a community setting. Sharon [Adams], at the right time, got up and directed when the time was right, she spoke most forcefully about what a big deal this was. If I was to go back and say what enabled me in this meeting, it was the inspirational leadership from Sharon. Then after it was all born, it became Marjorie [Mayfield Jackson], and certainly both exceeded my expectations by far. (Host 2012)

Mike Nickelsberg, retired Navy captain and academic biologist, who has served on the board's technical committee since the mid-1990s, coauthored the comparative risk assessment study on habitat loss, and currently serves as ERP's treasurer, said of Mayfield Jackson, the only person who has ever served as the ERP's executive director:

> You know why people stay with it . . . Marjorie and her vision. I mean . . . she's got this vision for the river and keeps the ERP focused on that vision and very much in a get-things-done type of thing. For a group like ERP . . . you need a leader with a vision. Not only vision but the energy to carry out that vision.(Nickelsberg 2012)

Bray, retired executive director of the Virginia Port Authority, characterized her in this way: "She is persistent, not obnoxious, but she stays in the game" (Bray 2012). Joe Rieger (2012), ERP deputy director for restoration, says of Mayfield Jackson, "Marjorie gives the vision. We couldn't be successful without that. She knows how to present the organization to the public and the media." Pam Boatwright (2012a), ERP deputy director for administration and head of the River Star Businesses program, says, "Other people are great, she just seems critical."

Dr. Huggett, chair of environmental sciences at the Virginia Institute for Marine Science, was another convener in the formative years. He was a recognized expert on the effects of contaminants on the river's fish population and became a member of the ERP's small governing board and project team in 1992 that developed the comparative risk assessment process. It was Dr. Huggett, a member of the science advisory board for the EPA, who went to Washington with Mayfield Jackson in 1992 to meet with the administrator of the EPA's Comparative Risk Analysis Branch,

Office of Policy Planning and Evaluation, which resulted in a small grant to begin the process. The ERP's founders gave substantial credit to Dr. Huggett for his counsel and guidance on scientific matters and credited him with keeping them grounded in real science (Jackson 2012b; Adams 2012a).

Later, there would be others who would serve as conveners, particularly the three presidents who followed Adams in that role. Retired admiral Ray Moses (NOAA) became the second president of ERP in 1995. His presence in that position provided heft for the organization in terms of access to and credibility with governments and business. As an admiral with NOAA he had worked with and met socially leaders of local and state government, as well as federal agency heads, and knew the industrial leaders along the river. Retired colonel Joseph Thomas, who as district engineer in Norfolk for the Army Corps of Engineers (USACE) when it began to cooperate with the ERP, joined the ERP board after retirement. He became its third president and during that period the Money Point cleanup, Paradise Creek Park, and Learning Barge projects were started. Money Point in particular required many months of negotiation headed by Thomas and Mayfield Jackson. Jim Oliver (2012), former city manager of Norfolk and then Portsmouth, said of these two:

> A huge piece was Ray Moses. I mean, Ray brought such credibility. He was as passionate as Marjorie was. He was so highly thought of. People like me just took a deep breath. Ray Moses, as the NOAA admiral, just puts a city manager at ease—Joe Thomas, from the Army Corps of Engineers, the same thing. Joe was president when I was there. Again, the evidence I saw at the retreats and board meetings, they clearly wanted every point of view to get expressed.

Another shift occurred during Thomas's presidency. The decision was made to change the makeup of the board away from its highly technical membership to include more community leaders who could establish a successful fundraising operation, and thereby decrease the organization's dependence on grants for covering day-to-day operations and projects. The next president, Bruce Bradley, former president and publisher of the *Virginian-Pilot*, led the organization into that territory (Thomas 2012; Bradley 2012).

Resources

The ERP founders had no financial resources when ERP was formed, nor did they appear to have prospects of finding funds to support their efforts. The founders got along without funding for much of the first two years. Most of what they did spend came out of their own pockets. The one exception was a small grant of $1,200 from the Virginia Environmental Endowment (Jackson 2012b). The budget has grown as has the dollar amount of the grants. The organization, however, has fed at the trough of grants for most of its existence. The staff and board understand that for an organization with an annual budget in excess of $1 million such a one-dimensional funding basis needs to be diversified. During 1996–1997, the first budget year after the Watershed Action Plan was implemented, 81 percent of the budget was from grants. That percentage was as low as 53 percent in budget year 2010–2011, though it is back to 75 percent in 2011–2012. The percentage changes from year to year are related to the types of projects ERP is implementing during the year (Elizabeth River Project 2012d).

Recently the board has made a conscious effort to shift its financial footing by bringing in more board members who have fundraising experience. Most board members spoke about the recent efforts to improve the fundraising capability during interviews and the need to restructure financial support. Current president Bruce Bradley and vice president Chris Neikirk reflect this change. As a result, fundraising activity has increased though grants remain the chief source of funding and ERP continues to be very successful at this.

The ERP has successfully leveraged its resources by creating partnerships in the watershed. Perhaps the best example of this is the River Star Businesses program started in 1997 with businesses located in the watershed, most on the river, the first such program undertaken by ERP. In 2012 the ERP partnered with more than ninety River Star Businesses, which conduct pollution prevention as well as wildlife habitat and mentoring programs (Elizabeth River Project 2012). More recently, partnerships have been established with the cities of Portsmouth, Chesapeake, and Norfolk, with businesses and nonprofits included in the partnerships. An example is the partnership between the ERP, City of Portsmouth, and Virginia Port Authority to develop the forty-acre Paradise Creek Nature Park project of restored forest and wetlands on Paradise Creek, in the middle of what was, until cleanup of several sites began some years ago, one of the watershed's most polluted industrial areas. The Paradise Creek Park project cost several million dollars and has received several six-figure contributions from diverse stakeholders. Another is the partnership between ERP, CBF, and the City of Norfolk to restore the Lafayette River.

Another source of funding has come through the Living River Restoration Trust (LRRT), established in 2004 by the Norfolk District of USACE and DEQ with a memorandum of understanding with the ERP that it will manage cleanup projects funded by the trust (Elizabeth River Restoration Trust Operating Agreement 2004). Its purpose is to use mitigation funds on projects in the Elizabeth River to offset harmful impacts from projects elsewhere on the river. The ERP contracts to manage the projects, primarily the removal of contaminated sediments from the bottom of the river. More than $5 million was received from APM Terminals Virginia as mitigation for the impacts to the river from constructing the APM Terminals in Portsmouth—on one of the river's cleaner sites. These partnerships and funding arrangements illustrate how ERP has evolved from a truly citizen-based collaborative to a mixed-group collaborative of private, public, and nonprofit partners.

A very different type of resource is the scientific and technical expertise the ERP has attracted from the beginning, without which it is doubtful ERP could have sustained its efforts and attracted so much national attention. Highly qualified scientists and other individuals with technical training and experience joined early and participated in the strategic planning process, later holding seats on the board and its committees. They made sure the ERP was firmly grounded in science in the early days and continue to take a hands-on approach to ERP plans and projects:

> We were very blessed by the good technical skills and the energy, goodwill, and commitment of the folks who participated. They brought depth, and real knowledge and understanding. (Adams 2012b)

It was not only the science from the academic community that ERP accessed, but the technical expertise of the regulator community. Frank Daniel, regional director of the Virginia DEQ, who became an active ERP board member after retirement, met with the ERP early and often in the 1990s. He characterized the relationship as follows:

> I became involved with the Elizabeth River Project very shortly after it was formed. Margie Mayfield approached me as the director of DEQ and told me that they were starting up the Elizabeth River Project and basically soliciting my interest and my help as a regional director of DEQ to help them get started and if possible to help give them insights into finding grant money through the state and at the federal level from the EPA. So that was initially how I became involved and that was fairly early on. I've given Margie advice about who to lobby and how to lobby in the state legislature. I also had numerous staff members who were involved and served on Elizabeth River Project committees, particularly their monitoring committee. We helped them quite a bit on monitoring water quality in the Elizabeth River. (Daniel 2012)

All of these efforts and many others could only occur because the organization took several years in its formative period to work on developing the trust among stakeholders who volunteered their time and expertise to develop programs, projects, and resources. That trust made it easier for local businesses and local, state, and national governments to make major contributions to ERP projects. The ability of Rieger, ERP deputy director, and David Koubsky, environmental project manager, to manage very large, but tight, budgets successfully, leveraging the available resources with in-kind contributions from contractors, while paying meticulous attention to detail at the site and completing the projects within budget and on time, has become a valuable resource for ERP and contributed to the trust that the broader local, state, and even national community has in this organization.

Rules

The ERP is organized as a 501(c)(3) corporation. As such it has a set of formal bylaws, though it operated informally as a citizen-based collaborative in the early days. The organization elaborated its structure and hierarchy as it evolved, most notably after the first Watershed Action Plan was completed in 1996, but it is still a grassroots collaborative, even though it has moved from being a purely citizen-based group to a mixed group of stakeholders from the public, private, and nonprofit sectors. Its meetings still tend to be informal and conversational in style and tone, which is not to say that they are inefficient. Rather than focusing on following formal procedure, members focus on solving the long-term and short-term issues by getting the best exchange of information possible. The real rules observed in the board meeting attended by the writer can be characterized as rules of civility, of hearing every person and recognizing every contribution. These were norms established as the basis of ERP operations very early in the organization's history, before it incorporated. They appear to have been a benchmark expectation since and a large contributor to ERP's success. Mayfield Jackson (2012b) described the procedures and rules followed in committee and board meetings as follows:

> We have an employee handbook and we have policies. There came a point where we had to be more formal. We had some growing pains on the way there. We seem to mostly have those in place to work smoothly. Sometimes people come on staff who have worked for larger organizations. We don't have a lot of red tape in the ERP. No. There are things in writing. Everything from conflict of interests to policies for the board. There are policies on who can speak for the organization. New board members get a one-page statement about board participation. Some are in the bylaws and some are not, the expectation that you give at least $500 a year, don't miss more than two board meetings, etc. Same thing with new staff. They get an employee handbook and sign off on various expectations. I think upfront that is very helpful.

On the other hand, the two-year comparative risk assessment process to develop the first Watershed Action Plan adhered to fairly strict guidelines. They had many meetings and retreats that were facilitated meetings. This is not an organization that leaves things to chance. One norm does appear to be that before entering into any new endeavor, the staff and board prepare thoroughly. Some might consider their approach to be unnecessarily slow, but that approach has been rewarded with successful projects. The ERP depends on those successes to maintain public credibility; the assumption is that every project must be a success. Anything else is not particularly tolerable. Approaching their work with that in mind, they are methodical in moving projects along.

Governance Structures

Mayfield Jackson was hired as a part-time staff person in 1992. A small office was secured in downtown Norfolk. After the Project Team of six finished interviewing sixty-five stakeholders, an advisory board was established. They used the following criteria to invite individuals from among those who had been interviewed:

- Persons who exhibited interest and enthusiasm about efforts to improve the appreciation and the conditions related to the Elizabeth River;
- Persons who because of their career or civic positions were positioned to understand issues related to the river and have access to a particular constituency;
- Persons who offered creative or positive visions when discussing the possibilities for the river (Elizabeth River Project 1993).

The advisory board met on September 10, 1992, at the Omni Hotel in downtown Norfolk to conduct its first organizational meeting. The governing board had by then been expanded to ten. Notable among the new members of the governing board was Dr. Huggett of the VIMS (Elizabeth River Project 1992). The expert on the effects of contamination on fish in the Elizabeth, he was the first scientist with ERP. It was Dr. Huggett who had the contacts at the EPA and helped secure EPA funding for the comparative risk process that would begin a year later.

The next several years were devoted to deepening trust and commitment to the process, conducting the comparative risk assessment, developing the strategies for addressing the highest risks, and concluding with the Watershed Action Plan. This process, described earlier, took the ERP through three major meetings. The 1993 Elizabeth River Visions meeting at the Waterside Marriott established the framework for the comparative risk process. The January 1995 meeting attended by 240 people, again at the Waterside Marriott, announced the results of the comparative risk process. The meeting at Nauticus in Norfolk, in June 1996, unveiled the Watershed Action Plan. During the intervening periods members were involved in many meetings and retreats as they worked to merge the diversity of opinions and interests into a unanimous consensus on all but one of the identified risks and then worked on task forces to develop strategies for managing the risks, which became the Watershed Action Plan.

The members had effectively spent the first five years of the ERP in a process to develop trust and social capital, on the one hand, and to build and agree on a plan to clean the river through a very deliberate and carefully sequenced process, on the other. They concentrated on building consensus in the community about what should be done about the river. The formal structure during those years included a well-organized effort with a small governing board, a large advisory board, and committees established to develop the comparative risk assessment, risk-management plan, and Watershed Action Plan. The focus was on the process. The governing board's job was to manage that process.

When Admiral Moses became president in 1995, the ERP's small office in downtown Norfolk housed the one-and-a-half-person staff of Director Mayfield Jackson and Pam Boatwright, who had joined in 1994 to work one day a week. The office contained a desk, chair, and lamp. He searched about to help them locate modest office furniture and copying equipment (Moses 2012). That was soon to change, along with the structure of the organization.

Shortly after unveiling the Watershed Action Plan in 1996 the ERP began the process of creating more structure, including adding full-time employees to the staff and creating a board of directors with standing committees to oversee policy and develop the projects that the ERP was finally ready to embark on. That year, 1996, when the Watershed Action Plan unveiled for the public, appeared to be a tipping point for ERP (Adams 2012b).

> There were eleven articles in the paper. It was on the front page on the *Virginian-Pilot*. Within a few months we had a staff of eleven and had more projects than we could keep up with. (Jackson 2012a)

Board member Mike Host (2012) characterized the change this way:

> What I remember is there was a transition after the plan got out on the watershed. ERP continued to try to handle some of these projects pretty much in grassroots mode, trying to get volunteers to do stuff and it was a lot of stuff. Plan a wetlands here, do more there, but I think at some point there was a realization, a gradual realization that you can work on these things as a volunteer, but you can't really work on it just with whoever shows up. And while everybody's opinion counts, not everybody has the resources to take action of this magnitude. So there was a gradual transition where the ERP I think started doing more with stuff like . . . hiring experts to help plan things out. Then when we got to big projects, like the Money Point project, for example, you need heavy equipment, contractors, people who expect to be paid for their work, so you had this split where part continued on with this grassroots thing, and another part that started gearing itself up to take on bigger projects, and those big projects demanded either big grant money or big donors with big pockets, and I think things became a little less grassroots.

There are several standing committees in ERP (see figure 4.2). An executive committee, appointed by the board of directors, includes all officers of the board and one member elected at large. It serves as the main planning group and advises the executive director. The executive committee meets each month; the full board every second month. The executive committee has the authority to act for the board between board meetings on most important issues that need immediate attention. Exceptions to this are amending the bylaws, electing or terminating board members, hiring or firing the executive director, approving contracts over $5,000, and approving major programs. A public relations committee has planning and oversight responsibility for several important functions, including public relations, marketing, education, fundraising, and serving as liaison to elected officials—city councils, state legislature, and federal officials. Several important advising and oversight functions are included under this committee's jurisdiction. One function is to monitor relations with elected officials, in which role they provide advice and strategy and join with the staff in interactions or negotiations with government agencies.

Another function of the public relations committee is education, which has always been mentioned in reports as a goal and function of ERP, but tended to be seen as secondary to the aim of cleaning up the river. That aim was understood to require partnering with the ship repair and maritime industries on the river to engage in cleanup actions, not educating the public about the river. But in 2008 the board made public

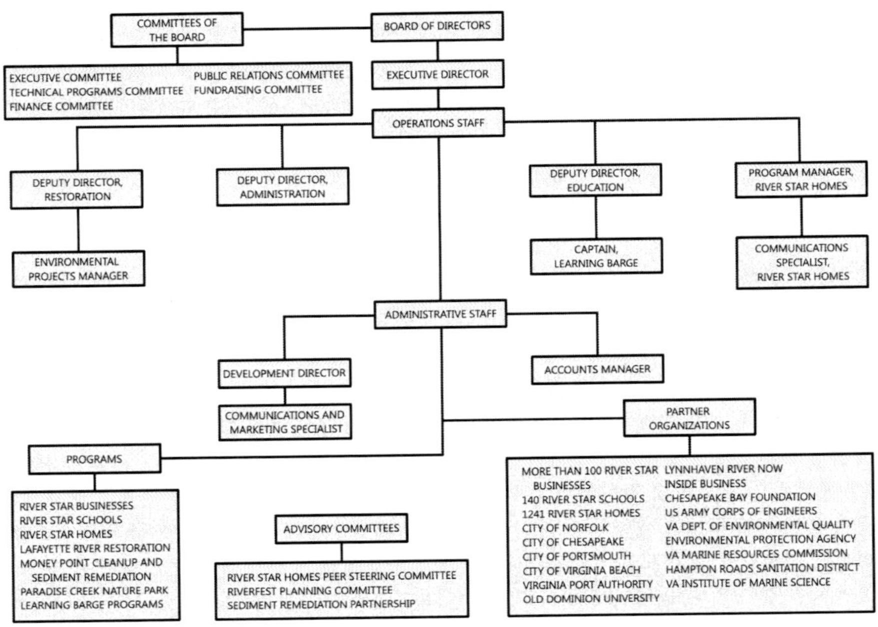

Figure 4.2. Organization Chart—Elizabeth River Project

awareness a high priority. This committee also supported the efforts of the fundraising committee. When Joseph Thomas became president in the early 2000s the board was still

> very technical and had a shortage of fundraisers. They were living pretty much on grants. The organization was very small. It was very technically based. That was about the time we realized that we needed to change the composition of the board. We needed to strike a better balance. We needed to actively recruit people who had the ability to help us raise funds. I think that was probably around 2003. (Thomas 2012)

For more than a decade, the board had been heavily weighted toward members with scientific and technical expertise; indeed, science is still well represented on the board. During the risk-assessment process in the early 1990s it was important for the science-based comparative risk reports to be grounded in good data and good science. Many scientists collaborated to provide it. Scientists on the board, most of them from two universities in the region with excellent programs in oceanography and marine science, VIMS and ODU, were instrumental in providing the solid scientific grounding that made it possible for the ERP to secure federal and state grants as well as achieve credibility with regulators and other stakeholders.

A transition in the role of the scientists and technical experts came after the First Watershed Action Plan was completed in 1996. As the ERP moved toward project implementation, members of the technical programs committee were charged with overseeing operational aspects of the technical programs. That includes assessing resource requirements for projects, monitoring achievements, reassessing program priorities, recommending new technical program initiatives to the board, and recommending policy positions if needed. Some were very involved in the design and implementation of projects (Jackson 2012c; J. Priest 2012). Although the ERP staff was adding personnel, it did not as yet have the technical and scientific expertise to manage the habitat restorations that were about to become identified with ERP. The technical programs committee more or less became the operational arm for those projects:

> Early on the board was very much involved in everyday operations, they would volunteer. In the beginning we weren't just board members, we were volunteer staff. The lines of communication got blurred. (W. Priest 2012a)

The project-implementation phase that ERP was moving into required similar scientific and technical grounding as had the risk-assessment and action-planning phase. This was what ERP had been building toward and the projects had to be successful. The ERP board and staff did not feel as if there was much room for error, so the technical programs committee and scientists on the board were the arbiters for the organization as it continued into the second generation of project development and implementation.

Several of the scientists who participated as early as 1992 and 1993 continue to serve on the ERP board and its technical programs committee today. Retired Navy captain Nickelsberg, former head of the biology department at Tidewater Community College, chaired the committee for many years before becoming ERP's treasurer. Dr. Joe Rule, retired associate dean of the College of Sciences at ODU, where he led a group in the early 1980s that conducted the most extensive research on chemical pollution in the sediments of the river, replaced him as committee chair. Dr. Michael Unger, research associate professor at VIMS, whose laboratory developed important measurement techniques for toxic chemicals in the river, became a board member and still serves on the technical committee. Walter Priest of VIMS, the region's most knowledgeable wetlands expert, participated since the first Marriott meeting in 1993. Priest left VIMS two years ago and moved to NOAA, from which the ERP has received several grants. Though he resigned from the board because of potential conflicts of interest, Priest remains very much involved in habitat restorations. He has participated in the design and construction of more than eighty acres of habitat restoration in over forty separate projects (W. Priest 2012b).

The technical committee has in recent years been augmented by retirees from state and federal regulatory agencies. Colonel Joseph Thomas, retired district engineer of the Army Corps of Engineers, served several years as ERP president from 2002 until 2010, during which the ERP undertook very large, multimillion dollar restoration projects, is a member. Bob Hume, retired chief of the regulatory branch of the Army Corps of Engineers' Norfolk District, joined the ERP and serves on the technical committee. Frank Daniel, retired regional director of the Virginia DEQ, joined the ERP and its technical committee after retiring. Hume and Daniel had been involved with the ERP process for several years as their agencies' chief regulators in the region and were familiar with the technical issues and the bureaucratic processes. These individuals and other scientist colleagues who participate give ERP's board a commanding scientific presence for developing and overseeing technical projects and building linkages to government and industry.

When ERP hired Joe Rieger, a biologist who had just completed a master's degree from ODU, to produce state-of-the-river documents and attempt to translate water-quality and sediment data into something accessible to the public, the board acquired a staff member with a science background. He quickly learned how to connect science with policy and projects (Jackson 2012c). His responsibilities increased and after a few years he began to manage the very large projects ERP was undertaking. Dave Koubsky, who had an engineering background, joined the staff to work on environmental projects. These two gave the ERP staff a scientific and technical edge that it lacked until Rieger arrived; this meant the staff now had the ability to manage operations. Indeed, these two have, according to many board members interviewed, given the ERP a strong internal capability to manage large, complex restorations and complete them on time and within budget. ERP has a reputation as a well-run organization as a result (Bradley 2012). This means the Technical Committee members now serve the more typical role of board member, advising, consulting, and approving policy and program plans, while the staff is entrusted to manage operations.

An ERP advisory committee is the Sediment Remediation Partnership Committee. It is "comprised of representatives of . . . federal, state and local agencies/institutions that have pledged to work together to . . . remediate contaminated sediments in the Elizabeth River" (Elizabeth River Project 2002). There was a need to bring together all agencies and institutions with regulatory authority and expertise to provide scientific and technical support on remedial action plans for ERP's main goal, removing sediments from the river. This committee was most important for the Money Point and Paradise Creek Nature Park restorations. The committee meets infrequently to provide guidance on the sediment removal issues.

Another recent shift in the board itself has been the addition of community leaders to develop its fundraising potential. The budget has been as much as $2.65 million in past years. Whereas dependency on grants has decreased from 81 percent of the budget in 1996–1997 to between 50–60 percent during the past four years, it fluctuates and is still too grant-dependent. A fundraising committee has recently been created, separated from the public relations–education committee, but it does not yet have a charter. The group is in the second of back-to-back capital campaigns (Bradley 2012). ERP vice president Chris Neikirk, vice president of Norfolk Southern Corporation, heads the ERP's fundraising efforts. As Bradley (2012) notes,

> One of our challenges is fundraising. It's just hard to raise all the money in a soft economy. We just got through an interview process for a grant and all of us agreed the biggest challenge we still have is being more proficient at fundraising.

Collaboration and the ERP: A Summary

The ERP began more than two decades ago as a purely citizens-based grassroots organization, with an agreement to not apportion blame for the problems of the Elizabeth River, understanding that its project would be doomed before it got off the ground if this were to occur. The members agreed that they would search for win-win solutions with all stakeholders involved in the decision process—all interests considered and protected. They offered a safe playing field for all. Even as they grew from a citizens-based group operating around a kitchen table to a mixed-group collaborative that includes stakeholder representatives from public, private, and nonprofit sectors, they have stuck to that strategy with a laser-like focus for more than two decades. All interviewed agreed that this has been the most important ingredient in ERP's success. It was how they encouraged stakeholders, who might have been expected to resist their ideas, to agree to continue the dialogue to the next level after the first interviews more than twenty years ago. Although it took several years for some of those stakeholders to develop substantial levels of trust, the process of creating social capital, a key ingredient for success, began by several accounts at the first meeting to kick off the comparative risk assessment process in 1993.

The Elizabeth River offered up such a complex set of problems—scientifically, politically, and economically—that it took a labor-intensive five-year process of collaboration to build the trust, and achieve the consensus, necessary to develop a Watershed Action Plan and move to the next level. That process was fueled with scientific research that balanced the science with what was politically feasible. Groups met and worked together for the first time. Attention was paid to the details of creating and maintaining a collaborative environment in meeting after meeting, by the founders, staff, and board. That consensus was achieved on a plan is a confirmation of the core values of ERP to be inclusive, make sure all voices are heard, approach partners as problems solvers, and never blame them for the river's problems. ERP would go on to create many enduring partnerships and implement many restoration projects, always based on principles of collaboration. The bigger, more complex, and costlier the project, the more meetings were held, the more stakeholders were involved, and the longer the decision process.

During the process the organization changed. James Thompson's (1967) work on organization examined the relationship between complex environments and organizational structure. He found that the more complex the environment, the more complex need be the organization. That is indeed what occurred with the ERP. From the grassroots beginnings and several years of what can be described as a somewhat loose, informal organization, the ERP went through a transition from a singular focus on strategic planning to an organization whose work was developing and implementing projects and building multiple partnerships to carry them forward. As it made that transition to a mixed-group collaborative the leaders found that more structure was needed to manage the more complex operations. The board built a structure of standing committees to oversee the process, and the staff increased to five times the original size from the planning years. The work of staff and board diverged, as the staff created the internal capacity to manage the complex technical projects and keep abreast of the science. The staff did the groundwork to develop ideas and proposals for projects. The board wrestled with the tough scientific and policy questions as well as how to best structure projects. The ERP took on characteristics of a "normal" organization in its structure. It was anything but normal, however, in the collaborative environment it created, the several partnerships it created and nurtured, the contributions made by unlikely partners, and the projects undertaken. The next section discusses those partnerships and projects.

COLLABORATION OUTPUTS AND WATERSHED OUTCOMES

The unveiling of the first Watershed Action Plan at Nauticus in Norfolk in 1996 marked a triumphal end to the first phase of the ERP. After five years of collaborative efforts to achieve the plan, ERP was ready to begin restoring the river in 1996. The first goal of the founders, getting the disparate group of stakeholders engaged in and committed to the process of collaborative problem solving with regard to the river, was successful. The first output was behind them; now it was time to move beyond that to the second goal of actually restoring the river. Thus began a series of programs, partnerships, and projects that continue today. The efforts to restore the river involve complex scientific calculations, substantial and complicated fundraising, dedicated management skill, and always collaborative relations among partners, since everything the ERP undertakes involves a partnership. We turn now to those restoration efforts and achievements.

Collaboration Outputs

One early effort—in fact, the first restoration project undertaken and completed by ERP, and the first volunteer wetlands restoration in the region—was not really a partnership, but more a one-man tour de force, or perhaps better described as a labor of love. That project, the Birdsong Wetlands on the Lafayette River by Larchmont Library, was named in memory of Dr. Ray Birdsong, ODU professor, who was the lead author of the habitat report for the comparative risk process in the early 1990s. It required two years to complete the restoration. Josh Priest, who grew up in the neighborhood and spent many hours on the river with his brother Walter as a child, managed the project, getting volunteer support wherever he could find it. He estimates that a few hundred people worked on it over a two-year period. He spent many nights over a period of several months working well after dark:

> When we did the Birdsong, I'd work from six to nine every night in the summer. My wife would bring the kids up to say goodnight. After about a year and a half nobody would come near me for fear I'd have them start working. I got used to being out there twenty hours a week because we ran out of money. (J. Priest 2012)

He estimates the cost to have been about $100,000 and thinks the contractor donated about $25,000 in design and construction costs. There were several civic league meetings where some citizens were skeptical about the project; convincing them took a considerable amount of effort and time. Four lots had to be rezoned. He had a different group of volunteers each weekend. Some weekends his "volunteers" worked because a judge had sentenced them to community service. He trained new volunteers on many of the weekends. ERP was important because funding agencies responded to its grant applications and contributed several thousand dollars to the project. The unpredictable occurred; he had planned to reuse all materials on the site. Then they found the bottles, thousands of buried bottles. Five dump truck loads a day were removed every day for a week; hundreds of cubic yards of material. He was left with a gaping hole and no funds to purchase fill. Fortunately, the US Navy had just dredged nearby and gave sand for the project. It was a learning experience, and it put a completed restoration on the map for ERP. Admiral Moses, ERP president at the time, commented, "To actually have something on the ground you could take people to see was very important" (Moses 2012). Asked how the site became degraded in the first place, Josh Priest told this story:

> It was—I don't know—nobody loved it. It was part of the original Larchmont plot that never sold because it was pretty low, and I guess nobody built there, then the library came along and it just sat there crumbling and we thought hey, let's fix this up. Something needed to happen and it was a good place to start and we learned a lot. That's the thing about doing an urban project like that; you never know what you're going to dig up. We did not do some exploratory digging. We found out we were going to dig up twenty thousand bottles. I did a presentation to a group of retired sanitarians and one guy said I was undoing his whole life's work. He was grinning so I knew he wasn't upset. He was telling me that in the forties and fifties they were filling in wetlands, so we were undoing his life's work. We found everything, the bottles, shoes, newspapers, signs, clothes, all kinds of stuff. (J. Priest 2012)

Restorations in the future would be undertaken by partnerships forged by the ERP, building on lessons learned, with organizations providing financial support and volunteers, most often through the River Star Businesses program ERP created that same year, 1997.

Plans, Agreements, and Partnerships

ERP's mission is to restore the Elizabeth River to the highest practical level of environmental quality thorough government, business, and community partnership. Now the ERP set about creating those partnerships necessary to restore the river. Industry along the river was, from the beginning, considered the main stakeholder.

After the Watershed Action Plan was launched in 1996, the staff began thinking about ERP's philosophy and history of bringing everyone to the table (Boatwright 2012a). They started Business Partners for a Cleaner River and Citizens for a Cleaner River. Very soon thereafter they went to the River Star Businesses model because they realized that it was too time consuming to get a couple of hundred square feet of restored habitat in a resident's yard, whereas with business there might be a couple of acres of restored habitat. ERP just didn't have the personnel to get to every citizen, and citizens who don't have much access to the industrial river are difficult to recruit. So River Star Businesses was launched in 1997 (Boatwright 2012a).

Boatwright, currently ERP's deputy director for administration, took over management of the River Star Businesses in 2000. By her account it was quite easy to become a River Star initially; the business simply had to commit to do some environmental stewardship projects. The problem was getting businesses to get involved with the program in the first place. "There was not a big commitment initially. It was a little bit of wariness of what is this" (Harper 2012b). The selling of River Stars was not characterized as a rational discussion of the benefits and costs, so much as selling a couple of businesses on joining (Boatwright 2012a). It was a new program with no advertising budget. Norfolk Boat Works, a small firm on the river, was first on board. It had shoreline erosion problems due to a storm water outfall and spent a small sum to create a shoreline buffer. Next, the Portsmouth Boating Center spent about $5,000 on a restoration (Boatwright 2012b). Once they had a few commitments and successful projects it became easier for other businesses to get involved. Boatwright still uses referrals from River Star Businesses, rather than relying on cold calls, to get access to potential new members. By 2012 there were ninety-two River Star Businesses (Elizabeth River Project 2012b).

The program is designed to get industries involved in two efforts: pollution prevention and habitat restoration. There are now three tiers and tightened requirements for membership. River Stars can begin at a "commitment level," for which the business has to document and quantify significant achievements in either pollution prevention or habitat restoration, with meaningful employee involvement. A second tier, the "achievement level," includes the same criteria as the commitment level, but requires both pollution prevention and habitat restoration. The third tier, the "model level," requires an additional component—mentoring other River Star participants. A river restoration advisory committee reviews applications each year for new membership and those wanting to move to the next level. Boatwright collects the information from the River Stars and writes a synopsis for the committee. It is first sent to DEQ and Hampton Roads Sanitation District (HRSD) for a compliance review. The advisory committee meets in the fall to approve a consent agenda, with recommendations from Boatwright (Boatwright 2012).

An applicant's achievements are recognized only when they go beyond regulatory requirements. Projects undertaken as compensation are not recognized; there has to be a net gain for the river. River Stars can renew their certification and keep their tier-rating by documenting maintenance of their responsible practices. ERP prepared a pollution-prevention workbook to assist River Stars—the P2 Workbook (Elizabeth River Project 1998)—and ERP staff meet with River Stars as needed to share ideas about what they can do to improve their environmental performance and advise them on implementation. Mayfield Jackson believes that everybody should receive a lot of public praise, an idea that has become ingrained in ERP's approach. Early each year ERP holds a recognition banquet for River Star Businesses, where many River Stars get recognition for their accomplishments during the previous year. In addition, a River Star Hall of Fame was created in 2010 to recognize one business each year that best exemplifies River Stars' commitments and achievements, using criteria developed by the river restoration advisory committee. The first three members are BAE Systems (2011), Virginia Port Authority (2012), and Kinder Morgan/Elizabeth River Terminals (2013). In 2011 *Inside Business: The Hampton Roads Business Journal* devoted the back page of the journal to River Star Businesses. It published an eight-page, $15,000 supplement on the accomplishments of Model Level River Stars in 2012 (Boatwright 2012b).

Boatwright (2012) says the "most successful River Stars come with top people wanting it to happen and a person in the trenches who has a fire in the belly, who can get it done on the ground." This describes the three Hall of Fame River Stars. BAE and Kinder Morgan/Elizabeth River Terminals, global corporations that had bought local companies, liked what ERP was doing. They gave the approval to work with ERP, thereby empowering local management (Ewing 2012a; Stedfast 2012). Michael Ewing and Phil Stedfast "had a fire in the belly" to undertake numerous costly restorations and pollution prevention projects (Boatwright 2012b). The Port Authority was rewarded for success in pollution reduction and restoring eleven acres of wetlands at Paradise Creek Nature Park. Heather Wood, director of environmental affairs, oversaw the Paradise Creek Nature Park wetlands restoration, and developed a "Green Operator" program to motivate and reward truckers entering the terminals to reduce pollution (Port of Virginia 2012; Mudflats 2012a; Wood 2012).

What motivates businesses to become a River Star? Several ideas were offered during interviews. Stedfast, manager of Kinder Morgan/Elizabeth River Terminals, a Model Level River Star and 2013 Hall of Fame recipient, said that while it is great recognition for the company, "It just makes me feel like I am making a worthwhile contribution" (Stedfast 2012). Kinder Morgan has a strong conservation and environmental ethic.

When senior executives at Kinder Morgan told him to pursue projects with ERP, he was able to put resources into the program (Stedfast 2012). He also suggested, as did others, that most businesses would prefer to be "painted with a 'green' brush." He offered the following with regard to the achievements of peers:

> It is infectious in that if your neighbor does something you kind of look over the fence and if they get recognized you begin to say that you should be doing that and we could do better. (Stedfast 2012)

The belief that peer influence had something to do with getting businesses involved was echoed by several other interviewees. Walter Priest agreed, but thought the most important thing that brought industry on board was the ERP's first decision—to not point fingers at industry and blame them for the river's problems (W. Priest 2012). He also echoed Ron Ritter's observation that there seems to have been a generational change that reflects a different attitude about environmental stewardship within industry (Ritter 2012). Ritter also thought it was an attractive tool for recruiting young employees. Admiral Moses, ERP's second president, like Walter Priest, believes the collaboration process itself motivates people (Moses 2012). Several people, as discussed previously, thought they detected this "buy in" to the process at the first meeting at the Marriott in 1993. The reality that regulators probably look favorably on River Stars is not lost on industry either. Finally, several interviewed suggested that the enthusiasm of River Star Businesses manager Boatwright is very difficult to ignore or turn away. Mike Ewing of BAE Systems said this about the ERP and businesses along the river:

> I think the Elizabeth River Project has done a good job of convincing business that this is a totally voluntary effort. It's the whole notion of encouraging businesses to go beyond compliance. And I think for new businesses coming into this they're still a little uncomfortable at first. They have to become real believers that, yes, this is really what that group is all about. They're not here to slap us around. I think that's what's made them so successful. You want to work with them. I think they have to be real careful about taking sides because I've been on both sides. I've been in academia, I've been on this side and I think some businesses would perceive that all you are is an extension of the EPA and so why would I be involved with your program. (Ewing 2012a)

Boatwright puts it this way: "We already have regulatory agencies. What we don't have is a collaborative carrot. It has been our defining characteristic" (Boatwright 2012a).

Ritter, who was interim president of Virginia Ship Repair Association when interviewed, said that when he was with Earle Industries, the company "would have met all those regulatory requirements, but would have met them without all the additional things that we did to protect the river" (Ritter 2012). This is one of the real contributions of the ERP. Businesses along the river often go above and beyond what is required by government regulations. They restore habitat that is not required by regulators, and they invest in pollution prevention above the requirements as well. Dr. Joe Rule, chair of the technical advisory committee, thinks that industry's "willingness to go beyond the state and federal requirements is absolutely remarkable" (Rule 2012).

BAE Systems Norfolk Ship Repair, the largest private ship repair yard on the river and first River Star Hall of Fame inductee, is an example of a business that has taken impressive measures to control runoff and restore habitat, some of them above and beyond regulatory requirements. Ewing, who heads all BAE environmental operations east of the Mississippi River, is the environmental manager there. When he arrived at Norshipco (now BAE) in 1999, after fifteen years at ODU, conditions in the work areas were such that workers at times had to walk through several inches of water and mud to get out of some buildings to the parking lots (Ewing 2012a). He had no expectations of saving the river, but did think there were several things that could be done to make a difference for the shipyard and river (Ewing 2012a). As had Stedfast, manager at Kinder Morgan/Elizabeth River Terminals, Ewing received the blessing of BAE Systems management to put energy and resources into making the shipyard cleaner and more environmentally friendly when it purchased Norshipco in 2005.[4]

The workforce, many employed there for as many as thirty years, had a longstanding way of doing things. It took some effort to get them to look differently at the environment. The "sell" was not as difficult as Ewing thought it would be, however. Many of those workers had grown up in the area and many loved to fish and would rather see clean water and air. They were aware of what they were doing when, after sandblasting a ship, they swept up what they could and sent the rest into the water. It didn't take as much persuasion as he had expected to get them to change their work practices. This is another example of how the attachment to one's place (BIMBY) creates commitment.

Two of the authors toured the facilities for three hours with Ewing to see the projects that many ship repair workers had volunteered for, including several wetland and upland habitat restorations, a three-quarter-acre urban forest on the river, construction of two oyster reefs, and raising thousands of oysters yearly to put onto the reefs. Two bioretention ponds were built for runoff from the industrial areas. The company also invested in building collection troughs around the dry docks to collect every bit of runoff water, and feeding it through a pipeline to storage tanks for treatment in BAE's water treatment plant. They tinkered with the wastewater treatment system until they improved it to the point where the copper levels in the treated water are below the levels in Norfolk's drinking water, thereby reducing the levels below what is required by the regulatory permit (Ewing 2012b, Rieger 2012). Tributyltin (TBT) is one of the deadliest chemicals used in antifouling paints for ships and is unusual in that it is so toxic that it is harmful to sea life in concentrations as low as a few parts per trillion (Unger 2012). Unger's laboratory at VIMS developed the technology to measure TBT at that level. EPA funded a five-year effort by BAE, Colonna's Shipyard, and ODU, to develop the treatment technology to actually remove TBT to the low parts per trillion (Ewing 2012; Boatwright 2012a). TBT is now banned in the United States, and an international ban on its use as an antifouling coating has recently been imposed. BAE's shoreline and oyster reef are featured on the cover of ERP's publication *Balancing Industry and the Environment: How to Achieve Win-Win on the Industrial Waterfront,* a primer prepared at the request of the EPA and funded by that agency (Elizabeth River Project 2008d).

Since the program's beginning in 1997, River Star Businesses have documented the following record for habitat restoration and pollution prevention:

- 1,194 acres of wildlife habitat created or conserved;
- 279 million pounds of pollution reduced;
- More than a billion pounds of other materials reduced and/or recycled (Elizabeth River Project 2012c).

As impressive as that record is, equally noteworthy and perhaps more important in the long run according to many interviewed for this study, are the behavioral changes in industry along the river, from the Norfolk Naval Shipyard at the mouth of the river to the River Stars in the southern branch. Scott Harper has covered the ERP for well over a decade for the *Virginian-Pilot* and had this to say: "This was the whole point for ERP, to build an ethic into those older industries" (Harper 2012). Ted Henifin (2012), general manager of the Hampton Roads Sanitation District and a major contributor to ERP projects, put it this way:

> I think changing the overall behaviors and attitudes is what's going to have an impact. So you've got someone like the ERP successfully convincing businesses to go in and do it on their own property, voluntarily, for a little recognition—that's an amazing feat that people don't really appreciate. But you've had shipyards and fairly dirty businesses decide to improve, take money out of their bottom line and stick it into their facilities on the ground and build, and it's like "how did they do that?"

Virtually every program and project is a partnership. The River Star Businesses program is the first and in some ways the most striking. They make a commitment to do habitat restoration or pollution prevention, or both, that goes beyond the legal requirements, and they fund it out of their own budgets. It requires them to go through an application process documenting their achievements. Once a commitment is made most sustain that commitment. It represents a significant turnaround from past industry attitudes and behaviors. That shift within the business and industrial sector may be the most important indicator of success for ERP, though that point is arguable since more recent projects and partnerships have taken the ERP in new directions and added more markers of success in cleaning the river. We turn now to some of those efforts.

Scientific Reports and Projects

ERP projects and reports, undertaken as collaborative partnerships, are too numerous to write about in this case, though each is important and deserves attention. Indeed, it is difficult for any ERP staff member to come up with a number or with documentation of everything undertaken by the organization since it ventured into ERP's second generation, project development and implementation, in 1997. The ERP website lists many publications and reports, as well as scientific studies, most conducted by scientists at ODU or VIMS. Virtually all efforts, studies, reports, and restorations undertaken by ERP are collaborative partnerships. In the previous section one partnership was highlighted, River Star Businesses. This section will focus on one of the many collaborative projects—Money Point.

Money Point is unique among ERP projects and is notable for several reasons. It exemplifies exactly what the founders discussed and identified as the main goal at the first meeting more than twenty years ago and has been uppermost on Mayfield Jackson's list since that time, which is to clean the contaminated sediments from the bottom of the river at the legacy pollution sites (Jackson 2012a; 2012b; 2012c). It has been the top priority in each version of the watershed action plan. Second, it is by far the biggest restoration for ERP; indeed one of the largest such cleanups on the Elizabeth. Third, it is the first cleanup that was not regulatory driven; several industries participated voluntarily. Fourth, it is the first large-scale sediment-removal project managed by a nonprofit grassroots environmental collaborative:

> What's amazing is that Money Point, that's the first cleanup that's not regulatory driven. That's a voluntary cleanup. And they're getting industry to do things not through regulatory channels, but through voluntary collaboration and cleanup and prevention. That's kind of—historically that's been unheard of. (Unger 2012)

The Money Point site on the southern branch in Chesapeake contained some of the highest levels of PAHs recorded anywhere in the world (Unger 2012). This toxic organic compound was responsible for the extraordinarily high levels of cancerous lesions occurring in fish in the vicinity of Money Point (Elizabeth River Project 2006). The PAH contamination resulted from the activities of several wood-treatment plants operating at Money Point until they finally shuttered their operations in the 1970s. Creosote, used to preserve utility poles, railroad ties, and bulkheading materials, leaked into the river over a period of several decades. A fire at the former Eppinger and Russell wood-treatment plant on Money Point in Chesapeake in 1963 released a large amount of creosote into the river, creating a thirty-five-acre biological "dead zone" laden with PAH (University of Virginia 2005). PAHs tend to settle into the sediment at the point of release, accounting for the high levels at the Money Point site. They diminish the benthic community that populates the river bottom and serves as the foundation for the food chain. The mummichog is a nonmigratory fish that lives in the marshy wetlands along the Elizabeth River and depends on that benthic community for its diet. These fish are highly sensitive to toxins, and since they don't stray more than several feet within their habitat, they are an indicator of the river's health. Those living in sites where wood-treatment plants deposited heavy concentrations of PAHs have severe liver cancers not found in other regions of the river (Unger 2012; Washington State Department of Natural Resources 2008). The mummichogs at Money Point have remarkably been found to alter their genetic code to protect themselves from the toxic diet. Without that genetic mutation, they would not live long enough to contract cancer. Mummichogs moved to Money Point from sites in the York River, where the river bottom habitat is healthier, died from contact with the toxic sediments there before they lived long enough to contract cancer (Harper 2010; Unger 2012).

When the APM Terminals were constructed across the river from the Lambert's Point coal piers, a vast amount of some of the healthiest river bottom was dredged. The USACE had used mitigation funds to establish trusts for wetlands restorations in other regions of the country, but the magnitude of the dredging at the terminal site was such that it required more than simply restoring wetlands and building oyster reefs. The amount of money available from the compensation paid by Maersk-ATM, more than $5 million, broadened the mitigation options. ERP's Watershed Action Plan for restoring the river called for removal of contaminated sediments as a top priority. It appeared that using the mitigation funds to remove contaminated sediments from the river might be the closest way possible to compensate for the damage caused at the dredged Maersk-APM site. EPA's support for ERP's Watershed Action Plan gave it scientific credibility from the Corps' point of view and thus presented the removal of contaminated sediment as a viable option for mitigating the effects of dredging at the new terminals. Hume of the USACE and Frank Daniel of the DEQ, the two regulatory overseers, agreed with the concept (Hume 2012).

The Elizabeth River Restoration Trust, soon changed to the Living River Restoration Trust (LRRT), was created with an operating agreement signed in May 2004 by DEQ, USACE, and Admiral Moses, chairman of the trust, to receive $5 million in compensation funds, as discussed in chapter 3 (Elizabeth River Restoration Trust Operating Agreement 2004). Moses was the link with ERP, where he had recently been president of the board. The trust was based on a model used previously by the Corps for wetlands mitigation, but never used for removing contaminated sediments and never on a project of this magnitude. DEQ and USACE agreed to let ERP manage the remediation project at Money Point using LRRT funds, though this was a move into new territory for the Corps. Joe Thomas, retired Norfolk district engineer for the Corps and president of ERP's board at the time, was instrumental in the negotiations between DEQ, USACE, and ERP because he still knew all of the officials at the Corps (Thomas 2012). His participation helped resolve the issues for the Corps. As for why a grassroots, nonprofit organization could be entrusted to manage such a complex project, respondents from all sectors had a similar answer. ERP staff had scientific and technical expertise required for the task; the board had deep scientific, technical, and management knowledge; and finally, the staff had proven themselves capable of effectively managing difficult collaborative projects. ERP's record for managing complex collaborative projects had established it at this point as a highly credible actor in decisions about the river's future.

The ERP's role was to contract with the LRRT to manage the project while the LRRT managed the funds. The money was available in 2004. Hume remembers that there were pressures to get something going with the project, but in 2008 there was still nothing on the ground at the site. "ERP's pushback was always, 'we need to do this right'" (Hume 2012). This time-consuming and methodical attention to the details was, Hume offered in hindsight, the right way, and he is glad for it. That approach has been characteristic of the ERP from its early days and must account for a measure of its success. ERP's drive to get the details right helps explain why ERP is authorized to manage a project to remove contaminated sediments, using funds from the very first trust established by the federal government for that purpose.

The restoration project managed by ERP is a collaborative partnership between ERP; the City of Chesapeake; the nine businesses located at Money Point (each a River Star Business); EPA; DEQ; USACE; and the citizens from the neighborhood at Money Point. A Money Point revitalization task force, comprised of almost one hundred participants from the partners, followed the time-tested ERP inclusive collaborative model, meeting for a year and a half, from early 2005 to late summer of 2006, to establish goals and develop the model for a "thriving waterfront industry that co-exists with ecological regeneration and yet affirms a residential community's history and safety" (Elizabeth River Project 2006a).

The plan called for an integrated restoration of sediment removal and onshore cleanup; wildlife habitat restoration that included tidal wetlands, riparian buffers, oyster reefs, upland forests; stormwater management improvements; and quality-of-life amenities for the local community. The first phase included removal of eight hundred cubic yards of sediment, roughly 750 tons, from the first toxic "hot spot," and replacing it with a "living cap" of clean sand, oyster reefs, and wetlands. Adjacent to this, five and a half acres of tidal wetlands and forested shore was restored on land that partner Elizabeth River Terminals owned and provided for the project (Elizabeth River Project 2009). The second phase, completed in 2011, removed more than fifteen thousand cubic yards of sediment, roughly fifteen thousand tons.

Funding came from several sources. The LRRT contributed $5 million from the APM Terminals compensation. The City of Chesapeake undertook $370,000 in stormwater improvements; there was no storm water management system previously at Money Point. The EPA contributed $1 million in the first phase for upland cleanup. The ERP contributed $177,000 secured with grants from the DEQ and NOAA. Hess Corporation provided $100,000 for the sediment feasibility study. The other businesses at Money Point contributed an estimated $3 million in pollution prevention, contamination cleanup, and habitat restoration during the several phases (Elizabeth River Project 2006b).

The first phase of Money Point was completed in 2009. All who were interviewed thought it would take years before aquatic life and water fowl returned to the restored area, yet before the first anniversary of Phase I the site was teeming with life. Walter Priest's (2012) observation, with regard to life returning to restored habitats, "If you build it they will come," was reaffirmed at Money Point. Seventeen species of fish were found there in 2010 (Harper 2010; Elizabeth River Project 2012e). First measures indicated that cancers in fish were decreasing. Professors Wolfgang Vogelbein and Michael Unger of VIMS, the scientists who have done most of the scientific research on the fish in this stretch of the river during the past two decades, found the precancer rate went down from 40 percent in 2006 at the Phase I site, to 6.6 percent in 2010, the year after Phase I was completed. This is almost at the background level of 5 percent, the level scientists consider normal (Unger 2012). This was a stunning affirmation of the recuperative powers of the river. It validated the efforts of ERP staff, board members, and partners. Mayfield Jackson considers the Money Point project to be the most important project undertaken to date by ERP:

> Because every iteration of our action plan has stated the top priority as cleaning up the river bottom, the toxins in the sediment and that is the problem that has to be addressed for the rest of the systems to recover and Money Point is where we made the significant in-road on that. (Jackson 2012b)

Conversations about sediment removal during the mid-1990s tended to be about whether it would be best to simply leave the contaminants at the bottom, since it was not clear if removing them would simply disperse them throughout the river, thereby making conditions worse. Even during discussions about the Money Point project less than ten years ago, capping the contaminated sediment with sand was a serious option. Improved technologies developed in recent years for skirting the dredge area, as well as new techniques for sampling and measuring contaminant levels in real time while dredging to see if PAHs are being released into the water column, made sediment removal a feasible option.

From ERP's perspective, Money Point is the jewel in Princess Elizabeth's crown. It provides evidence of real, measurable change in the health of the river. There is now tangible evidence, after years of successful collaborative projects, that the river can be improved by ERP's efforts. It had taken eighteen years to achieve this benchmark. Although restoring contaminated sediment sites is expensive, the toxic hot spots in the Elizabeth can be cleaned up when and if funding is available.

Education and Awareness

It is clear that all ERP projects and programs are partnerships. It is the case as well that education takes place as a result of all ERP programs and projects, whether intended or not. With that said, education has been considered something of a stepchild throughout most of ERP's twenty-one-year history. Virtually every person interviewed, when asked about the role of education, replied that it was of the highest priority. Yet on the scale of priorities it remained far below cleaning up the bottom of the river, as at Money Point, and engaging business and industry along the river to undertake pollution prevention and habitat restoration, as for the River Star Businesses. Nickelsberg (2012) said ERP needed to establish credibility before it could become an educator. Robin Dunbar, best known as Princess Elizabeth, is ERP's deputy director for education. It is an understatement to say that she is passionate about her job and about ERP. She is aware, as are other members of the board and staff, that education has been a stepchild at ERP (Dunbar 2012; Rieger 2012; Thomas 2012). By 2008 the board's position on priorities had changed:

> In 2008 the board's highest priority, besides implementing the Watershed Action Plan, was to increase public awareness. In the early years we just thought the main thing was to clean the river up and we didn't worry too much about education. But when we set the goal of "swimmable and fishable by 2020," the board and various advisers said you're not going to get there if more people don't know who you are and how they can personally be involved. So out of that came some really big projects: the Learning Barge, Paradise Creek Nature Park, River Star Homes. (Jackson 2012c)

River Star Schools The first education program was River Star Schools. Started in the late 1990s, there were five River Star Schools when

Dunbar arrived in 2000. Today there are 140 River Star Schools, which is roughly 60 percent of all schools in the watershed. Modeled on River Star Businesses, the schools engage students in projects such as oyster gardening, wetland plantings, litter cleanups, recycling, and water-quality testing for recognition at the achievement level. To reach model level, they have to go above and beyond those efforts and serve more than one year. The qualifications are not as rigorous as for River Star Businesses, which require empirical verification and peer review. Dunbar's position on this is as follows.

> I didn't make it so hard. When I adopted the program there was a committee that had to bless every project the schools did—and schools had to turn in this huge report. I got rid of that. These are kids, and a teacher that didn't have to do this, but somehow found a moment of time to say hey, let's do something related to the environment and the river. They should be recognized. (Dunbar 2012)

An example of one of these opportunities for young school kids is the Kinder Morgan/Elizabeth River Terminals program that buses five hundred kids a year to Money Point to transplant native trees and grasses that the kids have sprouted at school (Stedfast 2012).

The Learning Barge Oliver (2012) says, "I would say the ERP is about learning. In many ways it is about transforming and changing culture." That is Dunbar's mission. River Star Schools is a labor-intensive effort for Dunbar, who spends days, evenings, and many weekends at what might be described as a labor of love. It became even more labor intensive when the ERP took ownership of the Learning Barge in 2009, which quickly became the public face for ERP, creating much broader public awareness for the organization than had any of its prior efforts. The brainchild of Phoebe Crisman, professor at the University of Virginia School of Architecture, and designed by her students, the 120-foot vessel is a floating ecological, green classroom. It contains six learning stations that give young students "hands-on" experience to get in touch with and learn about a living marine ecosystem. In the Learning Barge's first few years, more than seventeen thousand students from kindergarten through grade 12 visited the Learning Barge and observed the relationship between the natural environment and a working industrial environment on the river (Elizabeth River Project 2012f). Working boats pass by on the river, reminding students of the connections between the natural environment and the world of industry, as students work at the interactive stations learning about human impacts on the river ecosystem, sustainable practices, tidal ecology, and environmental stewardship:

> It is getting kids exposed to planting a marsh or raising oysters. It is getting them to understand that stuff—that whole environmental education argument and getting kids out of the classroom and learning. You start getting teachers excited about training opportunities and get-

> ting the kids out of the classroom. I know there are a number of teachers active in ERP and LRN and they got involved in that way . . . again it is that process and getting people to think about the river and how they live. It's the long-term change. It is the best, I think. If you don't do that you only have the short-term gain. You won't have the lasting gain. (Harper 2012b)

The Learning Barge was not an easy sell to the staff or board. Crisman ran the idea by staff, but initially received no positive response. It was not considered a fit for the ERP's mission and came with a very large price tag; to take on the barge meant assuming responsibility for a very expensive asset, with substantial operating costs requiring a line item in the annual budget that could potentially overwhelm ERP's annual budget. When Dunbar met Crisman several months later and learned of her proposal, she went to Mayfield Jackson to ask why it was not approved. Mayfield Jackson's response was to go ahead if she thought she could make the project happen. A board that was by and large skeptical had to be convinced and $1.4 million would have to be raised. The idea was allowed to percolate for a few months until a majority of board members decided to support the acquisition, then a vote was taken (Dunbar 2012). Funding came from many nonprofit and private partners, as well as government agencies. The lead sponsor is Dominion Power, but there is a long list of contributing partners sustaining the barge. The maritime industries have assumed responsibility for maintenance. The list of maritime maintenance partners includes twenty-six River Star Businesses. Norfolk Tug Co. took responsibility for moving the barge during 2012. BAE Systems established a line item in its budget for barge maintenance. The Learning Barge is costly and therefore adds to the urgency to develop successful fundraising efforts, but it has paid handsomely in raising public awareness about ERP and the Elizabeth River.

Lafayette River Restoration Project A third component of education and public awareness developed along the Lafayette River in Norfolk, where several projects merged in the past three years to educate citizens about the river and to clean it up. The problems in the Lafayette, mostly a residential river, are high bacteria levels and stormwater runoff that create algae blooms on as much as 80 percent of the river in hot summer months, thereby depleting dissolved oxygen required for healthy aquatic life.

In 2009 the ERP joined in a partnership with the CBF to form the Lafayette River Restoration Project. At the time most of the Lafayette was unsafe for swimming and harvesting oysters because of high bacteria levels. The Lafayette River Steering Committee began work in March 2009, and worked for more than a year, convening about one hundred individuals from all sectors of the community—science, business, government, nonprofits, and citizens—in the ERP's time-tested, consensus-building process to develop goals and strategies. The plan, presented in April 2011, called for making the river swimmable and open to limited oyster harvesting by 2014, and safe for open oyster harvesting by 2020 (Elizabeth River Project 2011b). It outlined the actions that residents and stakeholders need to take to reduce bacteria and nitrogen runoff. The city of Norfolk joined as a partner and committed to continue its investment in stormwater management upgrades to prevent stormwater from overriding the sewer systems in heavy storms. The city has invested more than $76 million in this effort already. ERP is installing fifteen "scoop the poop" stations in the neighborhoods, working with the Norfolk Zoo on several projects, and beginning an effort to get the seven marinas on the river to join Virginia's Clean Marinas program. CBF is installing three hundred storm-drain medallions on storm drains in the watershed as a reminder to people to be mindful of what they put down their drains (Elizabeth River Project 2011b). Rain gardens, wetlands restorations, and tree-planting projects are underway by several partners, to complement several restorations already completed on the river. The ERP/CBF partnership has recently completed the largest oyster reef on the river (Elizabeth River Project 2011b). Testing for bacteria levels in the Lafayette was expanded to fifteen sites. CBF trained volunteers to be river stewards. In 2012 NOAA awarded more than $430,000 for oyster reef construction and wetlands restoration projects on the Lafayette by ERP, CBF, and the City of Norfolk (Chesapeake Bay Foundation 2012b; Harper 2012e).

Three days after the Restoration Plan was unveiled, the ERP held the first annual Lafayette RiverFest on April 30, 2011, on the Colonial Place waterfront in Norfolk. It was a festive occasion for the roughly 1,500 people who attended, with music, food, booths, workshops, and many opportunities to participate in "hands-on activities" to learn about environmental stewardship. The most important event of the day was the introduction of the new River Star Homes program, launched with the help of a $103,000 grant from the Hampton Roads Sanitation District (Mudflats 2012b). After twenty years on the Elizabeth River, the ERP was moving inland, into the neighborhoods beyond the shoreline to broaden public awareness and engage households and individuals in the effort to clean the river. The organizers thought they might get a few homes to sign up for the program during RiverFest. By the end of the day several hundred homes had signed up to become River Stars. Members are asked to follow seven steps to make lifestyle changes that will reduce the pollution running into the river. Participants promise to scoop up pet wastes, reduce the use of lawn fertilizers, keep lawn clippings and leaves out of storm drains and grease out of their sinks, not feed migratory geese, not flush medicines down the toilet, and pump out boat sewage at pump out stations (Elizabeth River Project 2012g; Elizabeth River Project 2012h).

In early October 2011, the project received a $301,900 federal grant from the National Fish and Wildlife Foundation (Harper 2011c). This allowed the ERP to provide financial aid to offset the costs of participating. These included discounted rain barrels, free lawn-care consultations and discounted "green" applications, and funding to help with shoreline enhancements and rain garden construction. It also helped the ERP hire additional help to manage the program and secure consulting advice for increasing membership and member involvement. The second annual RiverFest was another resounding success, in spite of rain and chill. By the end of 2012, a year and a half after the program launched, the program had 1,241 homes in the program, far above the two-year goal of five hundred, according to River Star Homes Program Manager John McLeod (2012).

Have the programs on the Lafayette been successful beyond raising public awareness of the watershed and its problems? Mayfield Jackson presented the results for three goals of a swimmable river, oyster harvesting, and reduced algae growth on a Lafayette River Scorecard in April 2012 at the second annual RiverFest. The data were quite striking. The DEQ found no bacteria violations of state safe-swimming standards in 2011. The Virginia Department of Health recorded that more than 50 percent of the river met bacteria standards for eating oysters. Neither means that one should swim in the river or eat its oysters as yet, but the improvements are impressive, and the change is in the right direction. While this cannot be attributed to ERP's River Star Homes—the stormwater and sewage infrastructure upgrades by the City of Norfolk and HRSD account for most of the progress—ERP's programs on the Lafayette are an important piece of the overall effort to clean this river, especially if progress is to be made on reducing algae blooms in the hot summer months:

> Without basic lifestyle changes among coastal residents, scientists and environmentalists say that restoring waterways, including the Chesapeake Bay, will be nearly impossible, regardless of how much money government spends. (Harper 2012c)

Paradise Creek Nature Park A final education program is the Paradise Creek Nature Park that opened on December 28, 2012. It is the effort of yet another extensive partnership comprised of ERP, many businesses, the City of Portsmouth, and government agencies. The forty-acre site is Portsmouth's second-largest public park; it provides public access to the river and features a $4 million, eleven-acre restored wetlands, a project contributed and managed by the Virginia Port Authority that is the largest restored wetlands in the area. Its canoe and hiking trails will provide an education forum that the Portsmouth school system will use, and for which its public-school teachers are already preparing lesson plans. It is also one of the two costliest programs, by a magnitude of several times, in the ERP portfolio (the other is Money Point). We discuss this with other education projects because it, unlike Money Point, is a project primarily for public educational and recreational use. It gets citizens close to a friendly river that is surrounded by heavy industry and otherwise would simply be inaccessible to the public. It is the only place where citizens can see a restored site on the river. It will alter public awareness of the river as it educates citizens about the river.

The project began as "Paradise Found" in 2002 with a grant from EPA funds administered by the CBP's Small Watershed Program (Elizabeth River Project 2003). Paradise Found was an initiative to restore all of Paradise Creek within five years. The "one creek at a time" model was ERP's idea for restoring urban watersheds on a scale amenable to collaboration. Whereas larger watersheds, such as the Chesapeake Bay, are too large to accommodate all stakeholders in a consensus process for short-term renewal efforts, a small 2.9-square-mile watershed such as Paradise Creek appeared to be an ideal site for enlisting multiple stakeholders in a manageable process. The scale made it possible to "take advantage of the small target area to ask virtually every homeowner, industry and government interest to do their part" (Elizabeth River Project 2008b). This attention to scale appears to be an important element in successful collaborations, and may be important for understanding why large-scale, statewide, or multistate collaborations seem to fail.

A stakeholder team of about fifty individuals representing all stakeholder interests on Paradise Creek—civic leagues; local universities; local, state, and federal agencies; the Navy; and industries—met in 2003 and reached consensus on a watershed action plan (Elizabeth River Project 2003). The US Navy became the first major partner with the ERP on the project. It was under a federal directive to clean up Superfund landfills on three sites along Paradise Creek where the Norfolk Naval Shipyard had dumped sandblast from its operations during World War I and World War II. The Navy was interested in community involvement on its projects. One, at the New Gosport Navy housing site, had already been cleaned up. The ERP proposed added amenities to the cleanup on another site in the 2003 watershed plan and the Navy responded:

> The Navy has an annex called Scott Center and that is located in between Victory Boulevard and George Washington Highway. We were thinking we can effectively cap this stuff and reroute the fence line. I talked to Marjorie Jackson. Mike Host scheduled the meeting and Mike and I went to talk to Marjorie about this site. They knew about it and knew about the good things we had done up there. She said, "You know, is there any way you guys can look at this site and do something similar that you did at Gosport?" So I looked at the numbers and crunched it. We ran some more tests. It was economical to remove it all, versus leave it in place, when you look at the overall life cycle. So in that case they really did bring the Navy in to reevaluating all options and reopening that feasibility study to incorporate their remedies. (Reisch 2012)

The industries along the creek took on new restoration and pollution projects. The Southeastern Public Service Authority plant received ISO 14001 international certification during the five-year period. Government agencies undertook environmental enhancement projects.[5] These activities led to the Paradise Creek Nature Park project. ERP purchased the property along Paradise Creek in 2007 for $1.2 million, with funding from the Virginia Land Conservation Foundation, Virginia Port Authority, and the *Virginian-Pilot* (Elizabeth River Project 2008b). Other sponsors joined in the project that will cost $15 million when the final phase is completed in 2014. The largest grants of more than $500,000 were contributed by the City of Portsmouth, which will manage the park and incur an estimated $500,000 in expenses over the next several decades; the Virginia Port Authority, which spent $4 million to construct an eleven-acre wetlands in the park; the Virginia Land Conservation Foundation; and the Virginia Department of Conservation and Recreation (DCR). At least six others contributed from $100,000 to $500,000. Many individuals and organizations have made smaller, though substantial, contributions.

The ERP staff, Rieger and Koubsky, are again managing a major project for the ERP, and as with Money Point, have engaged contractors who have a successful track record with these types of environmental projects and are committed to working with the staff to achieve the best outcomes and reduce costs wherever possible. Rieger and Koubsky have themselves now established a track record of efficient and effective management of large-scale environmental restorations, as a team that can complete them on time and under budget, and the budgets have always been tight enough that there is no margin for error:

> One of the things we do when we do these big projects is we go through a very competitive process review of proposals to make sure we are hiring someone who is experienced and also take in the factor of cost. When you do that, you save a lot of money on the back end. I feel it is my job and duty to make sure we get the best product out of it. I don't want to spend extra money that someone has sent to us or from a grant on something that could have been prevented. Dave and I, when we do these projects, we really spend a lot of time looking at innovative ways to reduce costs and working with the contractors to make sure they feel vested in the project and they like the product that they are providing and that at the same time we are getting the best product that we can. We don't hand a project to someone else to manage where they have to come up to speed on the project. We are the ones that do everything from work with the concepts to the design, to bidding to construction to monitoring. (Rieger 2012)

Koubsky and Rieger also developed social capital with workers at the site. Koubsky builds goodwill with simple acts of kindness such as bringing work crews watermelon on hot summer days.

The project is yet another example of ERP efforts to include neighborhoods in the collaborative planning process. The neighborhood of Craddock, across the street from the park, was involved from the beginning in the process, according to Craddock Civic League president Dottie Wyatt (Wyatt 2012). Teens from the neighborhood planted native vegetation in the park. Koubsky began working in early 2012 with a Craddock youth group and he started a Park Ambassadors Program for youth from the neighborhood (Koubsky 2012). Portsmouth school teachers are now writing lesson plans incorporating the park into the science curriculum.

Watershed Outcomes

The most notable outcome is the return of many species of fish to Money Point and a remarkable reduction in cancers in the mummichog population less than a year after completing Phase I in 2009. Seventeen species of fish and shellfish were found in the restored marshland there. While not a direct measure of cause and effect, it is persuasive evidence of the quick recuperative powers of the river.

A second outcome attributable to ERP is the behavioral change observed among River Star Businesses. Ninety-two have instituted pollution-prevention and habitat-restoration projects that go beyond regulatory requirements and at considerable cost that none were required to assume. Many of those interviewed for this study talked about how this change in behavior occurred because ERP was there providing justification for their efforts, creating a vehicle for their participation (River Star Industries), offering suggestions and guidance, and cheerleading. Their position is that this additional investment of time and human resources by River Star Industries simply would not have occurred without ERP (Harper 2012; Unger 2012; Host 2012; Bray 2012; Henefin 2012; Ewing 2012; Rule 2012).

Other measurable improvements in the Lafayette River, significant increases in oyster population and water-quality improvement, have occurred where ERP has been involved with restoring habitat, building oyster reefs, and educating homeowners about eco-friendly lawn and pet care. While these cannot be linked directly to ERP actions, their work appears to be an important piece of the overall cleanup efforts on the Lafayette, especially toward creating sustainable behavioral change among residents.

SOCIAL CAPITAL

Bingham, O'Leary, and Carlson (2008, 6) point out that collaboration goes beyond cooperating to developing new structure, sharing resources, defining relationships, and communicating. It involves creating and building social and organizational capital. The ERP is an exceptional study of how a group of four individuals with a vision, but no material resources, can enlist the participation of often-wary multisector stakeholders in a long-term collaborative effort to solve "wicked" problems and cultivate deep levels of social capital in the process. They began as a citizen-based group that reached out to all sectors which, by virtue of the nature or place of their work—business, government, military, science, and civic groups—were connected to the river. They appealed to their sense of the historical, cultural, and economic significance of the Elizabeth River, and most responded with meaningful contributions of time, resources, and ideas. Many who participated as volunteers in the early 1990s are still with ERP in 2013.

The founders understood at the very first meeting that to move forward with their plan to clean the river, they could not assign blame and would have to include all stakeholders in an open and transparent collaboration process. Virtually every person interviewed attributed the ERP's longevity and success to these two norms. Though the founders brought no material resources, they did bring considerable social capital themselves. Their respective professional experiences, as well as their work in environmental organizing and policy advocacy, gave them the skills, self-confidence, and trust to believe that the democratic, collaborative process they envisioned to clean the river could be successful. Yet, the scenario presented a daunting environment for the founders. The stakeholders operated in a sometimes hostile climate where the values of business and profit often clashed with the environmental regulatory actions of governments. They had not necessarily, in previous encounters with one another, established the norms of trust, reciprocity, or legitimacy with regard to one another that are required for an effective long-term, collaborative partnership. Many were wary of becoming involved initially, yet stayed with the process set in motion by ERP and made appreciable contributions, even becoming vocal cheerleaders for ERP. The success of ERP in building social capital among such a diverse group over the first several years and sustaining the participation of stakeholders over the long run is a testament to the power of collaboration.

The first step in the process, reaching out to interview sixty-five individuals in 1992, was counted a success (Adams 2012; Heinricht 2012). They were not thrown out of any interviews, but instead left each with an individual willing to continue with the project. With each step in the five-year "process" leading to the first real project, the Birdsong Wetlands in 1997, the social capital developed among the large group of stakeholders deepened and set the stage for the many successful projects undertaken later. ERP president Bruce Bradley (2012) said the decision to never "point fingers" to blame a party was the most important decision the organization ever made. Trust developed among participants as they worked together on difficult issues. The process gained legitimacy during the first large meeting at the Waterside Marriott in 1993. Many current members who participated in that event saw it as a seminal moment for the organization because such disparate stakeholder groups sat and talked and listened carefully and learned about one another. Cheryl Barnett (2012), environmental director for the Norfolk Naval Station in the 1990s, member of the first Watershed Action Team and ERP board member in the 1990s, thinks it all goes back to the inclusive way ERP began at that meeting and has remained faithful to since. Wagner and Fernandez-Gimenez (2008) found that the respect and understanding that develops when diverse stakeholders in environmental collaboratives talk and listen respectfully to other points of view, an essential aspect of ERP's process, was essential to success in the several environmental collaboratives they studied. They also note that predictability and dependability are essential to promote social capital. ERP has been absolutely predictable in following its norm not to "point fingers" of blame. Mike Host, former environmental manager for the Norfolk Naval Shipyard and ERP board member, thinks that

> it would not have been possible if the ERP had not established itself as an entity that could be absolutely trusted, over and over again, all those relationships built from 1993 on, all those meetings where everyone got to be heard, all those balanced decisions. All of that stuff led to a point where the state and the EPA and city governments and all those guys could say, ok, we trust you . . . the trust is huge, absolutely huge. (Host 2012)

Rieger (2012) described ERP's ability to get industries along the river to talk with an environmental group about the Money Point cleanup this way:

> First what we did was we had a group called the sediment remediation partnership. It is one of our technical committees that has partners on it. We knew Hess was contaminated so we have always included them on updates of other projects.[6] We didn't always focus on Money Point. We discussed other sites. Over time the Hess representatives started to interact more with our staff and understand more about our organization. Slowly we started building that relationship. In 2005 we said, "Let's look at Money Point for a cleanup project." They helped us fund the feasibility study. They didn't insert their bias if they had one. In my opinion they saw it as a way to move the project forward, be cost effective, and have the ability to work with us on the final outcome. They had seen other projects in the watershed that were just the opposite. They remediated the uplands through the volunteer remediation program, but the offshore is an issue no one wanted to talk about. We don't care who did it; we just want to come up with a way to clean it up in this lifetime. Hess realized that, and we worked together. I think they are happy with the product so far. You have to take time to get to know each other and then you start developing trust. Building those

> relationships is so important. Contractors work with you if you are open with them. Everyone puts their cards on the table. That is why it has been so successful. As a staff person you have to work hard at keeping those relationships, not only managing projects.

Ron Ritter, interim president of the Virginia Ship Repair Association and longtime executive in the maritime industry on the river, said this about ERP's relationship with industry:

> We're not environmentalists, you know. We're ship repairers. So it's nice to be able to pick up the phone and call somebody like ERP who isn't going to put you on a report with the Environmental Protection Agency to say, "I think I have a problem, could you come over and help me? I think I may not be doing this exactly right, can you come and talk about what I might do differently?" And they brought in an environmental consulting group, Williamsburg Environmental Group, which I thought did a wonderful job and at a very reasonable cost to provide their expertise through ERP. I would never have found them without ERP. (Ritter 2012)

Social capital is developed with industry workers as well as work crews at construction sites. Ewing, environmental administrator for BAE Systems, has overseen many restorations on BAE property. He made the following observation about the shipyard workers:

> If you're asking for volunteers [for a restoration project], we would probably get dozens of people that would be willing to show up on a Saturday on their own time and help. They would feel like they're contributing somewhere. I think these guys know . . . these guys have been around here for thirty years . . . every day they can feel the difference when coming to work at a cleaner facility where we're trying to protect the water and the air and the critters and everything else. (Ewing 2012)

Tim Reisch, manager of several Navy Superfund site cleanups, said about workers at one of the worst, the Atlantic Wood site, where ERP was involved as a member of the shipyard restoration advisory committee:

> The Atlantic Wood site was done in 2003. We excavated this all out. We had to bring in a tremendous amount of fill and create the channels. I will tell you the workers absolutely loved working on that site. What accounts for that? They got a sense of accomplishment. It was a site with twelve or thirteen feet of calcium hydroxide sludge and you have truck load after truck load taking stuff out. They participated in it and they really felt some accomplishment. They made a difference. (Reisch 2012)

ERP began as a purely citizen-based collaborative.[7] It has been characterized in this chapter as a mixed-group collaborative, at least since 1996. Bidwell and Ryan (2006) and Bonnell and Koontz (2007) report a higher level of conflict in mixed-group collaboratives, attributed to the considerable diversity among stakeholders, than in purely citizen-based or agency-led collaboratives. We conclude that ERP's success at engaging such a diverse set of stakeholders in the difficult consensus-based process to accomplish so much is a testament to the power of social capital to reduce conflict in mixed-group collaboratives.

Commitment to Place

Commitment to place has always been a strong motivator for participation with ERP. The founders recognized the potential power of this attachment as they prepared to conduct interviews with stakeholders in 1992. "We used the argument of history, culture, and pride in the river, that those who participated could be keepers of that tradition" (Adams 2012a). Each of the four founders had their own particular commitment to the river, though none was originally from the area. They had advocated for the river and organized award-winning, annual trash collections along the shorelines. Mayfield Jackson actually lived on the river, at Scott's Creek in Portsmouth, after arriving in the area in 1983. She described her experience this way:

> People thought the river was pretty much dead. They didn't even know its name and the headlines in the paper were that the fish had cancer. I had taken a sabbatical from the paper for six months so I had spent a lot of time sitting in the backyard. I would see herons eating fish. Clearly it was not dead. It was quite beautiful in my little corner. When I would walk out on the dock, I would try not to look down. I was ashamed that I lived in a community where the river had such toxins the fish had cancer. That is really why I tried to do something about it. At some point I decided what would really matter to me was if I could help clean up that river that was in my backyard. (Jackson 2012a)

Walter and Josh Priest, both very active contributors to some of the best-known ERP projects, exemplify the powerful motivating force growing up on the river can have. Both expressed joy when talking about their childhood in the Larchmont section of Norfolk. The love they express for the river is best illustrated for Walter by the fact that he became the wetlands restoration guru for Hampton Roads, the person called on when a restoration plan is evolving. Many interviewed told how he could "see" when looking at a damaged site; he understands the potential and what needs to be done to realize it (Reisch 2012). He has often arrived at a site, studied it, and taken out a pad of paper and pencil to sketch the design for a final restoration model. His brother Josh felt so close to the river and the neighborhood of his youth that he spent a year and a half there doing the first volunteer wetlands restoration in the watershed, at the Birdsong Wetlands, often working at night for weeks at a time. Most of those interviewed had lived on the river, worked on the river, regulated the river, or conducted scientific research on the river and spoke of the attachments and sense of responsibility that generated. This reflects the "paradox of scale," discussed by Bonnell and Koontz (2007), where local-level watershed problems tend to have greater salience for stakeholders. It is easier to commit to solving watershed problems in one's own backyard, indeed many became involved "because it's in my backyard."

SOME FINAL THOUGHTS

The Elizabeth River Project has accomplished much in twenty-one years. Of that there can be little doubt. Many lessons can be drawn from the case; some of the most important are listed in textbox 4.1. For all that has been achieved—getting into offices to talk to stakeholders about the river, conducting years of collaborative meetings to develop a watershed action plan, developing and implementing programs and projects to move the action plan forward, and finally completing the large restoration projects—there is unanimous agreement among all who took time to share their thoughts with the study team that the major reason for success is the collaborative process itself. It was the willingness to always engage all stakeholders in an open process and never try to apportion blame. The opposite side of that coin is the culture of recognition and praise ERP nurtured, where all participants get recognized publicly for any achievement.

Textbox 4.1

Lessons from Elizabeth River Project

- Conveners with appropriate skill sets, commitment, and energy are important for launching successful collaborative processes.
- Successful collaboration requires inclusiveness and transparency at each step. It helps build social capital that is necessary for success.
- Publicly recognize the contributions of particular stakeholders.
- Prepare thoroughly before bringing diverse stakeholders together. Take the time necessary to lay the groundwork, rather than moving quickly in response to pressure from the public or stakeholders. This may include beginning the consensus-building process before a first meeting.

- Achieve success as soon as possible. Getting diverse stakeholders together to talk and listen to one another can be an important success and begin the process of building social capital, as well as a motivator for continued participation.
- A competent facilitator is an important element for successful meetings with diverse stakeholders.
- Adopt a behavioral norm to not blame others for the problems.
- Commitment to place is an important motivator for participation and success.
- Participating stakeholders from particular sectors bring other participants from the same sector into the collaborative process.
- Success brings more participants into the process.
- Collaboration itself can be a powerful driving force for participating and is often the strongest "glue" for continued stakeholder participation.
- Collaboration helps leverage and broaden available resources.
- Never neglect to pay careful attention to the "details" of the collaboration process itself.
- Successful early collaborative meetings and completion of a first successful project builds social capital, creates public credibility, and brings other stakeholders into the process.
- Scale is important.
- Education builds committed citizens for future success.
- Stakeholders will be determined by the nature of the problems.
- The availability of scientific and technical expertise is important for building credibility and assuring the integrity of projects.
- Utilize the expertise of retired persons from the public, business, and academic sectors where possible.
- The collaboration process is the single most important driving force for success.

They achieved several firsts for a grassroots environmental collaborative. There were no models when the four individuals met in 1991. Environmental collaboratives were relatively new then and they were generally agency-based, led by government agencies. Although they have grown into a mixed collaborative of partners from government, nonprofit, and industry, ERP began as a citizen-based group and helped develop the grassroots model as an alternative form of environmental organization. The Money Point restoration is another example of a "first." The dredging and capping done there was not unique; it was the process that was unique and broke new ground. It was done outside the regulatory process as a voluntary project with a fully engaged community of stakeholders from start to finish. Voluntary process and full community engagement are the unique contributions at Money Point. ERP had built that into its model in the beginning.

What is remarkable is that so many individuals and organizations have been so thoroughly engaged with ERP for many years. They focused on technical restoration projects and the collaborative approach used to make them successful, on developing broad networks of River Stars to build social capital, create a connection to the river, and get them involved in developing and implementing their own projects. Interestingly, when discussing the big accomplishments, the most often mentioned was getting industry to think differently about the river and change its behavior toward it, and getting trust and support from so many diverse stakeholders and citizens. McCarthy (2012) cited creating awareness of the river as something that needs to be restored, as the most important accomplishment. Reisch and the Priest brothers said it was cheerleading public awareness (Reisch 2012; W. Priest 2012; J. Priest 2012). Adams (2012b) said it was getting people connected to the river. Rieger (2012), the project manager for the big projects, said it was community engagement. Mayfield Jackson, the unwavering leader with that laser-like focus on cleaning up the bottom of the river, said it was a paradigm shift in industry after a tentative start (Jackson 2012b). Many of those who have participated with ERP in the past two decades appear to agree that educating stakeholders and the public about the river, leading the behavioral change among stakeholders, and deepening their connection to the river is the organization's most important legacy that creates the necessary foundation for sustaining the cleanup of the river.

NOTES

1. Toxic legacy pollutants are chemicals that remain in the environment long after they were produced. The legacy pollutant sites in the Elizabeth River were mainly deposited before the era of modern environmental regulation.
2. Adams and Mayfield Jackson recounted the amount of the grant as $1,200.
3. Nauticus is a maritime-themed science and technology center located on the Elizabeth River in downtown Norfolk, Virginia. It includes the Hampton Roads Naval Museum.
4. Norshipco was Norfolk Shipbuilding & Drydock Corporation until purchased by BAE Systems in 2005.
5. NOAA funded a half-acre oyster reef; the City of Portsmouth worked with ERP to establish the largest rain garden in the region, funded by the Virginia Department of Conservation and Recreation; DEQ started a water-quality- and sediment-monitoring program on the creek; USACE began studying how to address contamination in the creek's sediments; the EPA got involved in regulatory action to clean up PCBs at a private site.
6. Hess Corporation operates a petroleum terminal at Money Point.
7. See chapter 2 for a discussion of three types of collaboratives.

FIVE

The Return of the Oyster to the Lynnhaven River

I have never encountered anyone that says they do not believe in what we are doing. Culture change doesn't happen because somebody in Washington tells us we have to change our culture. It happens because people start to care.

—Karen Forget (2012d)

On April 21, 2012, Lynnhaven River NOW (LRN) celebrated its tenth anniversary with its annual Oyster Roast celebration and fundraising event. This was no small gathering. Five hundred tickets were sold to the event, which was held on the Lynnhaven River under a huge tent on the front lawn belonging to Bob and Eleanor Stanton. Bob Stanton was one of the founding members of LRN and has graciously hosted the Oyster Roast each year. Anyone at the event would have noticed right away a truck bearing the logo of the Lynnhaven Oyster Company. The Lynnhaven Oyster Company was one of the caterers of the event and provided many of the oysters for the five hundred guests. The oysters came directly from the Lynnhaven River and are said to be some of the finest eating oysters in the United States. The oysters for the event were donated by the watermen working the Lynnhaven River, which included Hap and Cameron Chalmers of the Lynnhaven Oyster Company and Chris Ludford and Chris Pace. There were no watermen harvesting oysters from the Lynnhaven River in 2002 when LRN was formed, since the river at that time was too polluted to allow oyster harvesting. The tenth anniversary celebration of LRN was a remarkable demonstration of the power of a grassroots collaborative effort, and eating oysters from the Lynnhaven River was proof of their accomplishments.

There were many other organizations and individuals that came together to make the tenth anniversary Oyster Roast a success, and many of these organizations and individuals have been involved with LRN right from the beginning of the organization. Even the weather cooperated so that the event could be held under a sunny sky. It was impressive to see the number of people who volunteered their time to make the event such an important achievement. A decade ago it would have been hard to imagine that an event like the Oyster Roast could even have been possible in 2012 because the Lynnhaven River was in such bad shape.

THE CONTEXT AND HISTORY OF LYNNHAVEN RIVER NOW

The Physical Setting

The Lynnhaven River is located entirely within the city of Virginia Beach in southeastern Virginia. The Lynnhaven, with its connected Broad Bay and Linkhorn Bay, is the largest and most developed river system in Virginia Beach. The river encompasses sixty-four square miles and nearly 250,000 residents live within its watershed. The three branches of the Lynnhaven have over 150 miles of shoreline and hundreds of acres of marsh, mudflat, and shallow water habitats (see figure 5.1).

The Lynnhaven watershed drains north to the Chesapeake Bay and is subject to the ebb and flow of the tides. Lynnhaven Bay is a trap estuary

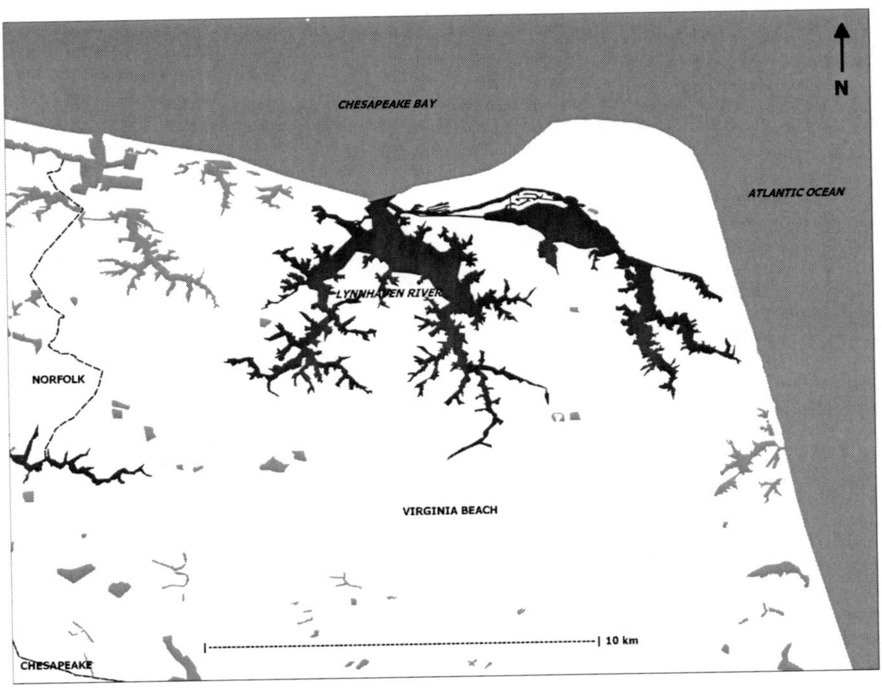

Figure 5.1. The Lynnhaven River Watershed

that receives freshwater input from coastal streams and urban runoff. However, the system receives little freshwater inflow from its small coastal tributaries, and its circulation is dominated by tidal flushing with waters of the Chesapeake Bay through the narrow Lynnhaven inlet constricted by the Lesner Bridge. The Lynnhaven is a shallow system with a maximum depth of nine meters at its mouth.

The area comprising the Lynnhaven watershed can be characterized as highly urbanized and densely populated with a high proportion of residential properties. An estimated forty thousand dogs reside within the Lynnhaven River watershed. Undeveloped land comprises less than 25 percent of the total watershed as forest, wetland, urban grassland, or water (DEQ 2004).

Karen Forget, writing in a book titled *The Lynnhaven: Restoring a Legend*, which was a product of LRN, noted that:

> The river has an impressive history! Some of the earliest European settlement in our country was along the Western Branch of the Lynnhaven. Historic homes and important sites abound. Four hundred years ago, the river was the main thoroughfare and a major source of food for these early settlers. World renowned oysters, seasonal delicacies like shad roe and soft shell crabs, plentiful blue crabs and a wide variety of fish provided some part of almost every meal. Throughout the years, many have made their living on the river harvesting this bounty. (Forget 2011, 4)

A little over four hundred years ago on the shores of the Lynnhaven River, the first permanent English settlers came upon a band of Indians roasting oysters, which the settlers described as large and delicious in taste. In the heyday of the Lynnhaven oyster in the early 1900s, they were a favorite of gourmand Diamond Jim Brady. President William Howard Taft dined on the famous Lynnhaven oysters on a visit to O'Keefe's Oyster House in old Princess Anne County (now part of Virginia Beach) and ate so many that he said he felt like an oyster (Barrow 2004). The secret to the Lynnhaven oyster's flavor was always said to be its wonderfully salty taste caused by the river's proximity to the Atlantic Ocean. In the stores, according to Mary Reed Barrow, small oysters are sold as "Standards" and large oysters are sold as "Selects." Once, many years ago, Lynnhaven oysters were so big that they needed a description unto themselves. The Lynnhaven oysters were sold as "Counts" (Barrow 2004).

The City of Virginia Beach was a small, independent city in the 1950s. In 1963 the City of Virginia Beach and Princess Anne County were consolidated as a new, much larger independent city, retaining the better-known name of the resort town along the Atlantic Ocean. The city experienced rapid growth in the 1970s and 1980s, and the population is now more than 420,000. The city is primarily residential, but it has a large military presence with a master jet base for the Navy.

Political Culture

Daniel Elazar (1984) identifies the dominant political culture in Hampton Roads as traditionalistic/individualistic. While there are elements of these cultures in many of the cities, Virginia Beach in 2012 is much different than it was when Elazar first developed his typology in the 1960s. First, the city has experienced significant population growth. Moreover, most of the population growth is due to an increased military presence in the region; the population today is a true melting pot of people from all parts of the nation. This has significantly weakened the traditionalistic attributes present in the city forty years ago.

Second, the city has developed a highly professionalized city-manager government. The current city manager, James Spore, has served as city manager in Virginia Beach for more than twenty years (Spore 2012). Spore has endeavored to provide a wider range of city services, and has done so with the support of both the city council and the citizenry. While there is still an active libertarian citizen contingent in the city, the culture appears to be dominated by moralistic characteristics. Elazar (1984, 117) suggests a moralistic culture that:

> emphasizes the commonwealth conception as the basis for democratic government . . . both the general public and the politicians conceive of politics as a public activity centered on some notion of the public good and properly devoted to the advancement of the public interest. Good government, then, is measured by the degree to which it promotes the public good . . . and commitment to the public welfare.

Commitment to the public welfare in Virginia Beach is illustrated, in part, by the willingness of the city to support the activities of LRN in a very direct manner.

The State of the Lynnhaven River in 2002

In 2002 when LRN was formed (the organization was originally named Lynnhaven River 2007), the Lynnhaven River had been almost completely closed to oystering for decades due to pollution. Sections of the river were closed to shellfishing in 1930, and the main stem of the Lynnhaven was shuttered for good by the State of Virginia in 1986 due to bacteria pollution. The oyster industry in the river was extinct shortly thereafter. Swift waterfront development in Virginia Beach during the 1960s and 1970s was the main culprit. The city population grew at a very rapid rate and with that growth came hundreds of pleasure boats, restaurants, and shopping malls. Prior to 1990, development in the Lynnhaven watershed was basically unregulated in terms of water quality, and since the watershed is primarily a flat coastal plain, there were always flooding issues that contributed to the river's pollution. The stormwater system was designed to drain water off the land as quickly as possible. There were grading requirements for lots so that water would run down to the beach or the river behind the building—whatever was the quickest way to get the water off the land. Even to this day there are still approximately 1,200 pipes that dump stormwater directly into the river every time it rains, and each of these pipes has several intakes (Forget 2012a).

The Lynnhaven River is not polluted due to industry, factories, or power plants; there are none of these on the river banks. The river's pollution is caused by ordinary people's oil, gas, antifreeze, pet waste, car-wash soap, lawn fertilizer, pesticides, septic-tank leaks, and all manner of crud washing untreated into the river, what regulations call "nonpoint source" pollution (*Virginian-Pilot* 2003). In 2002, when LRN was formed, there was only 1 percent of the river open to any shellfish harvesting. As several people noted, the Lynnhaven River had been "loved to death."

Formation of the Organization

It all began with a conversation between two men in 2002 and their vision for the river. Andrew Fine was the CEO of the Runnymede Corporation, which was involved in real estate development in Virginia Beach. Harry Lester was the president of the Eastern Virginia Medical School. Both men, long-time friends, had worked on a large capital campaign to raise money for the Chrysler Museum in Norfolk. The campaign was very successful, and as Fine put it, they just asked themselves, "OK, what should we do next?" The answer was, "Why don't we tackle the Lynnhaven River?" (Fine 2012). Fine had grown up around the Lynnhaven River and had spent a great deal of time on the river in his youth and had fond memories of taking oysters out of the river, making a fire, and grilling them right on the banks of the river. Lester lived on the river and had strong attachments to it. Almost immediately these two men included Bob Stanton in their plans for the river. Stanton is Lester's neighbor on the river and his brother-in-law, and also had a long history of time on the river and an equally strong attachment to it. Stanton is the president of Stanton Partners, Inc., which specializes in commercial real estate investments. By any measure, this is an unusual group of men, two of whom are involved in real estate development, to establish a nonprofit organization dedicated to the environmental pursuit of cleaning up a river. Their vision was simple and compelling. They wanted to be able to eat oysters harvested from the Lynnhaven River by the year 2007. The vision was simple, but the task was not. The state of the Lynnhaven, as detailed above, was deplorable with significant pollution and no particular culprit at which to point a finger.

One of the first meetings initiated by Fine and Lester was with the city manager of Virginia Beach, Jim Spore. As Spore (2012) recalls:

> We were sitting around having lunch and they say, "We have this idea." What is it? They say they want to turn the quality of the water around in Lynnhaven River and we have decided we have got to get something really symbolic that people can identify with. We want to be able to eat oysters out of the Lynnhaven River by 2007. It's the 400th anniversary of the landing at Cape Henry. So I said to them, "Have you guys lost your minds?" They said they thought they could do it. They said, "We think we have to have a visible goal that people can really understand and relate to, rather than talking about all of the esoteric stuff in terms of water quality. The litmus test is going to be the oyster and the ability to eat the oyster in the Lynnhaven River." That was the initial meeting.

The result of the meeting with Spore was a proposal for a partnership between the city and the fledging nonprofit organization that the founders put together. The new nonprofit, Lynnhaven River 2007 (which will be referred to by its current name, Lynnhaven River NOW, or LRN, for the remainder of this chapter) needed resources in order to be a viable organization. As Stanton (2012) put it, "We went to the people who live on the Lynnhaven and asked them 'Would you rather have a clean river or a dirty river in your backyard?' It was fairly easy to build a fairly substantial base of financial support." The founders mostly recruited people they knew, initially between thirty and fifty people, and asked for a thousand dollars a year for three years, which gave the organization a financial base for operations. Many of those original contributors still support the organization with donations.

The founders set the tone for the new organization. One of the things the founders agreed was that LRN was going to be a voice of moderation. They were not going to attack anyone, but rather serve as facilitators for people and organizations that didn't have a great working relationship historically or no working relationship at all. As Fine (2012) noted, "We considered ourselves to be 'middle of the roaders' and we made that very clear. If you wanted to be a radical environmentalist this was not the place for you." At the beginning the founders sought advice from other organizations including the Elizabeth River Project, the US Army Corps of Engineers (USACE), and the Chesapeake Bay Foundation (CBF). They discovered there wasn't really a model for what they were trying to do. The Elizabeth River Project (ERP) faced a different set of challenges on the industrial Elizabeth River with major point-source pollution problems. The CBF told the founders that LRN was going to have to be the new "model" for cleaning up a residential river where the primary enemy was residential runoff.

The founders believed that the key to success for LRN was going to be a strong partnership with the City of Virginia Beach local government. As Fine (2012) put it, "If we hadn't gotten the positive reception we did from the City Manager from day one, I am not sure we would have proceeded." Many city departments and individuals were involved in the Lynnhaven cleanup effort. As Fine (2012) noted:

> At first we sensed some resistance from the departments who had historically been extraordinarily independent, but about a year into this thing, suddenly the choir started singing the song back to us and we realized that a lot of these people who had gone into professions with ideals, suddenly realized that they could do something that was really extraordinary. They could do this simply by energizing the forces that they had available to them and directed around this project. It was a very nice thing that happened.

The founders and the city manager understood that there were some things, like interacting with citizens, that a nonprofit organization, with committed and passionate members, could simply do better than city departments. Spore, the city manager, noted that:

> The real key to the Lynnhaven is going to be getting people to change the way they behave in terms of their lawn practices, animal practices, boating practices and all that kind of stuff. It's those nonpoint things at this stage of the game that we needed to address and we didn't have enough staff to address it. It was really gratuitous in terms of timing because we had gone about as far as we could with the typical public works fix. We really needed some partners and it was perfect timing. (Spore 2012)

Partnership would be the pathway to success in cleaning up the river.

The Executive Directors of Lynnhaven River NOW

In the ten-year history of LRN there have been three executive directors. The first director was a city employee, Barry Frankenfield. The founders asked for a "loaned executive" from the city to help get the new organization off the ground. This was an unusual arrangement, but it proved to be very successful. Frankenfield was working in the Parks and Recreation Department at the time. According to Spore, Barry had really good people skills and worked well in small-group meetings with citizens (Spore 2012). It required some effort and negotiations to figure out how the loan arrangement was going to work. In the end it was agreed that Frankenfield would work at another location two days a week as the first executive director of LRN. Frankenfield recalls that the organization began with almost nothing; some initial money had been raised by the founders, but there was no fundraising program and no staff support—he was on his own to get things moving (Frankenfield 2012). The founders did engage Dave Iwans and Associates to provide assistance with logo design and a template for brochures and publications and set up the organization's website. The work was done on a pro bono basis. One of the first orders of business for Frankenfield was to make sure that fundraising efforts provided enough of a base to pay his salary for his time at LRN. The agreement with the city called for LRN to pay the equivalent of his city salary for the two days a week that he was "employed" by LRN. This was actually easily accomplished through the fundraising work of the founding members of LRN. Frankenfield had other important priorities as the new organization got up and running. They were to develop specific projects for the organization to undertake, begin community outreach efforts, and examine pertinent regulations. The original office for LRN was in the Runnymede building owned by Fine and offered free of charge. As Frankenfield (2012) noted, the fledging organization received a lot of pro bono help from a number of different organizations and individuals.

Frankenfield served as the part-time executive director of LRN for two years. During that time the structure of the organization began to emerge with several different committees of volunteers taking responsibility for specific activities, such as education, oyster restoration, landscape, and governmental relations. The committee structure will be discussed in more detail in the next section of this chapter. The selection of a city employee as the first executive director was a major advantage for the organization since it went a long way toward ensuring a good working relationship, cooperation, and trust between LRN and the city departments involved in the river cleanup effort. City employees were active members on the various committees that did the real work of the organization, and their involvement on the committees was strongly supported by the city manager and his executive staff. Frankenfield, with the approval of the city manager, was able to facilitate a city contract with LRN to provide community education that was funded from the stormwater funds of the city. This funding partnership continues and provides a source of income for the organization. At the end of two years Frankenfield felt that the organization was financially stable and needed to get to the level of having a full-time executive director. It was time to move the organization forward. Reflecting on the origins of LRN, Frankenfield (2012) noted that in his opinion the founders were able to put together a group that led the city in a new direction in terms of dealing with the pollution of the Lynnhaven River watershed.

The second executive director of LRN was Laurie Sorabella. Laurie held a master's degree in marine science with a concentration in oyster reef and underwater grass restoration from the Virginia Institute of Marine Science (VIMS). Her involvement with LRN began as the education director for the organization. She had written a grant application that funded her position with the grant money coming from the National Fish and Wildlife Foundation. When Frankenfield was ready to go back to work full time for the city after two years with LRN, Sorabella stepped in as executive director and served in that position for three years. The organization was still young when she took over and still could not quite afford a full-time director, but she was working three-quarters time in the position. Under Frankenfield's leadership the organization already had some notable accomplishments such as the "Scoop the Poop" campaign and the initial organization of the committee structure. Sorabella moved the organization forward with education programs in the schools and by establishing working relationships with the watermen on the river. During her tenure as executive director the Lynnhaven River was designated as a "no discharge zone" for boaters.[1] This was a major accomplishment that required the cooperation of many different stakeholders. At that time LRN also took a position against the development of what is now known as Pleasure House Point. The property, directly on the river, had been purchased by a developer that planned to build 1,100 units and fill in many acres of wetlands. Opposing the development was a bit controversial and made LRN's executive committee a little nervous, according to Sorabella, but the development would have been devastating for water quality in the river (Sorabella 2012). The development was never built. The property is now owned by the city and will never be developed, although there are plans to build an environmentally friendly facility to house the headquarters of the CBF's regional office and the office of LRN.

Under Sorabella's leadership the committee structure expanded and more organizations, such as USACE, the Virginia Marine Resources Commission, and the CBF began to participate as members of the various committees. Sorabella resigned as executive director after serving for three years due to family and other obligations; however, she is still very active in LRN as the co-chair of the oyster restoration committee and in educational activities for the organization. She also runs the small nonprofit Oyster Reef Keepers of Virginia and consults with both ERP and the Nansemond River Preservation Alliance (NRPA).

The third and current executive director of LRN is Karen Forget (pronounced For-jáy), a former biology teacher and oyster gardener. Forget began work at LRN as the education and outreach coordinator and has served as the executive director for about five years. LRN has grown in size during Forget's tenure as executive director. There are currently four full-time staff and two other people who work part-time on a contract basis. The newest staff position is the restoration project coordinator, which gives the organization the capacity to work more directly with homeowners to really promote shoreline management and living shorelines. The organization also works on a contract basis with Oyster Reef Keepers, which operates the school system's oyster-gardening program as well as other related programs. LRN contracts out bookkeeping and website work, but still receives pro bono work from some other organizations, such as Dave Iwans and Associates. The annual budget for LRN runs around $400,000. Forget now works with a board of directors of nine members, including two of the original founders. There is also an executive program committee that is comprised of the chairs of each of the organization's committees. The LRN committee structure will be discussed in more detail in the next section of this chapter.

Funding for the organization has expanded during Forget's tenure. There are federal and state grants, and one of the education programs has been generously funded by the Dollar Tree Corporation for several years. There is a contract with the City of Virginia Beach Department of Public Works to do a stormwater education and awareness program. Other funds come from the Virginia Beach Oyster Heritage Fund. Donations and memberships are a significant part of the organization's income, and the Annual Oyster Roast is the major fundraising event each year. Forget (2012c) characterizes the Oyster Roast as a "friend raising" event. Forget also serves as the representative from LRN on the city's Clean Water Task Force, which is a committee made up of city employees and representatives from a number of nonprofits that coordinates programs in all of the different watersheds in the city.

Under Forget's leadership, the organization has continued to embody the collaborative spirit that LRN began with in 2002. From the very beginning, the philosophy was that nobody could do the cleanup alone and everyone would have to work together to be successful. Many key city employees, including the city manager, give Forget credit for the willingness of LRN to work in partnership with all the stakeholders on the river. As a testament to all the partners, the Lynnhaven River is now much less polluted than it was ten years ago when LRN was formed.

The Watermen on the Lynnhaven River

Commercial watermen on the Lynnhaven River were almost nonexistent in 2002 when LRN was formed. The river was simply too polluted to support any commercial shellfishing since the 1980s and most of the river hadn't been open for shellfish harvesting for decades before that. The older watermen on the river, such as N. B. Godfrey, Irvin Evans, and John W. Keeling, are long gone, but today there are a handful of oystermen working the river. With the increasing water quality in the Lynnhaven in the last several years and the return of grounds to commercial shellfish harvesting, there is now a market for Lynnhaven oysters, and that emerging market signaled the return of the watermen to the river.

Perhaps the first commercial oyster operation of the new era was the Lynnhaven Oyster Company, founded in 2005 by Hap Chalmers and Cliff Love. At that time between 10 and 20 percent of the river had been reopened to shellfish harvesting. The company really started as a hobby, a backbreaking hobby as Hap Chalmers noted. He further stated that:

> We just wanted to see if it could be done because it hadn't been done in about fifty years. We did a couple hundred thousand—and that was enough to prove to us that it could be done. We did not plan on actually continuing to do it. Cliff left the business and I did too, but then I

> came back this year [2012] to help my son financially to expand. (Chalmers 2012)

In the beginning, according to Cliff Love, Rob Brumbaugh was contacted for advice. Rob was the director of the Hampton Roads office of the CBF at the time and was invited to serve as a science advisor to the newly formed LRN. In Cliff's words:

> In 2003, I joined Lynnhaven River 2007. The organization had the ambitious goal of improving the water quality so that Lynnhaven oysters could be eaten safely by 2007. Rob Brumbaugh guaranteed that by then the water would be clean enough for him to eat an oyster. . . . With the help of all the members of Lynnhaven River 2007, the City of Virginia Beach, the Virginia Institute of Marine Science, the Virginia Marine Resource Commission and the Army Corps of Engineers, our river is getting much healthier. In 2007, Rob came back to the shores of the Lynnhaven and fulfilled his promise. (Love 2011, 7)

Hap's son, Cameron, now runs the Lynnhaven Oyster Company with five full-time employees. The company leases about five hundred acres of Lynnhaven River bottom, which is enough to sustain a good-sized commercial operation. The company is now producing three to four thousand oysters a week and hoping to produce thirty-five thousand a week in 2013.

The practice of oystermen has changed over the years. Chase Groseclose, a former LRN staffer, wrote that:

> Many of the oysters are grown using sustainable aquaculture techniques which maintain the native stock and increase the oysterman's efficiency. Cameron Chalmers, owner of Lynnhaven Oyster Company, grew up on the water, worked on a charter boat and has fished most of his life, but he never thought he would end up an oysterman. But Chalmers has been farm raising oysters full time since 2007. "It's by far the best job I've ever had!" he said. Because of the success of his operation along with those of the other Lynnhaven oystermen, the local treasures can be found in restaurants and seafood markets across the region. (Groseclose 2011, 55)

COLLABORATION PROCESS

From the very beginning, LRN was founded on a philosophy that emphasized collaboration, partnership, and civility. The founders understood that the restoration of the Lynnhaven River could not be accomplished without the collaborative effort of many different partners and stakeholders. There was no desire on the part of the founders to create an organization that was confrontational in nature. They essentially championed the idea of collaborative effort. Forget reflected on this in one of our interviews with her. She noted that:

> You need champions, you really do. If you can find some people who are already well respected in the community or are viewed as really good community members, then that goes a long way in working with the city council. They see this as a win-win kind of thing. The city has to be your partner. A nonprofit cannot do the kind of work that needs to be done to improve a waterway. We can guide, we can encourage and do our own projects and education and participate and be a partner in all kinds of efforts, but the local government has to be one of your major partners. They have to be or you are not going to be successful. You don't go to them just when you want them to do something for you. You have to be engaged with them. You have to really work with the city and be their champion too and really promote the efforts that they are doing and try to work with the citizens to better understand how the city government works and how we can work together. I think that is one of the things we have done really well, and that has benefited us. We have formed a very collaborative relationship with the city government and where they see this as a win-win effort and respect for what we are doing. We try to really work with them and help them in any way we can. That is not the way some nonprofits work, but to establish an antagonistic relationship with the city or county is stupid and foolish. (Forget 2012a)

In 2004 the Virginia Beach City Council passed a resolution that cleaning up the Lynnhaven River was a "highest priority" for the city. That resolution has gone a long way in helping strengthen the collaborative relationship between LRN and city staff in the effort to clean up the river (Forget 2012a). In interview after interview with members of LRN, the key to the organization's successes in all its various projects and efforts was identified as the ability to collaborate and partner with stakeholders. Steve Herbert, a deputy city manager with the City of Virginia Beach, referred to LRN as a "force multiplier." He noted that:

> We don't have the staff and resources to get that focused and that much energy devoted to something like the Lynnhaven River cleanup. LRN's performance has allowed us to really excel in an area that I don't think anybody would have assumed. We have made advances and progress in terms of water quality, but I don't think anybody would have imagined this happening. I am not sure we would have done this on our own. We would have eventually gotten around to fixing the sewers in Little Neck, but LRN has the benefit of organized passion, volunteers, citizens that are influential, and they have adopted an operating mode that has been very successful. (S. Herbert 2012)

Clay Bernick, in the City of Virginia Beach Planning Office, commenting on the collaboration between the city and LRN, noted that:

> The city wouldn't have had as much citizen support without LRN. I think that is where the big difference was. It can't be an issue focused group. I think that is what has been the strength of something like LRN. They are not focused on dredging in the Lynnhaven or fighting

> marina proposals. So many citizen organizations seem to be focused around a particular issue. I think LRN took another approach. The stewardship ethic is much better communicated through a community group than through the city government. (Bernick 2012)

Stakeholders

The nature of the Lynnhaven River watershed, which is located entirely within the city limits of Virginia Beach and is almost completely residential, with no industrial development, has strongly influenced who the primary stakeholders are in the effort to clean up the river. LRN began with a small group of individuals who shared a vision of a clean river that could produce a harvest of edible oysters by 2007. The founders clearly understood that engaging all stakeholders in the cleanup effort would be the only viable path to success. The initial effort to establish LRN focused on attracting a base of citizens, those living on or near the Lynnhaven River, who could provide financial support for the new non-profit organization. This was not a "hard sell" since there was a desire in the community to do something about the deplorable state of the river. Those residents living directly on the branches of the Lynnhaven River certainly had a very strong stake in the quality of the water, especially since a polluted river would drive down property values. However, all residents of the watershed had to be thought of as important stakeholders since the state of the river was a "quality of life" issue for the entire community. Programs such as "Scoop the Poop" could only be successful with wide community participation from all residents of the watershed, not just those living directly on the river. To expand their influence with residential stakeholders, LRN has been actively engaged in working with garden clubs and civic leagues in the Lynnhaven watershed. LRN has run educational programs with these groups to build awareness of landscaping practices that are river-friendly.

When LRN was formed in 2002 there were no commercial watermen making a living on the Lynnhaven River. There were several marinas and seafood restaurants on the river that would have to play a role in the cleanup efforts, especially by enforcing discharge rules for their boat owners and by operating "clean" marinas. Early on LRN reached out to the boating community and obtained its help in converting the Lynnhaven River into a "no discharge" zone. This regulation could not have taken effect without the support of this important group of stakeholders. As the river got cleaner, more of the river bottom was opened to shellfishing, which brought commercial watermen back to the river, and they became stakeholders in the river cleanup efforts. LRN developed a close working relationship with some of the watermen and had them serving on some of the committees so that they had a voice in the strategies and plans for the river.

One of the most important stakeholders is the city government of Virginia Beach. The city must be concerned about issues related to total maximum daily loads (TMDLs) in all of its watersheds. As described throughout this chapter, the City of Virginia Beach has developed a strong partnership with LRN, as well as other citizen-based organizations, through its Clean Water Task Force and Green Ribbon Commission. Based on its partnership with the city from the very beginning, LRN is a "mixed group" collaborative as described in chapter 2. Many city departments are directly involved in activities related to the quality of water in the city's watersheds, and a number of city employees serve on the various working committees of LRN. It would also be appropriate to include the city school system as a stakeholder in that it is responsible for teaching the next generation the ethic of environmental stewardship of the watershed.

The CBF, USACE, and a number of state agencies are also directly involved with the Lynnhaven River and are important stakeholders in the state of the river. The CBF is concerned with the larger region and the Chesapeake Bay as a whole, but its members have been very active in providing advice and resources to LRN and the city on issues related to the Lynnhaven River. The current plan is for the CBF to build its new Southern Bay headquarters on the Lynnhaven River. The Corps has committed significant sums of money to oyster-restoration efforts on the Lynnhaven River and maintains an active presence on LRN committees. State agencies, such as VIMS, have contributed technical expertise to LRN that has proven invaluable to the restoration efforts on the Lynnhaven River.

The majority of the Lynnhaven River watershed was developed with primarily residential properties in the 1960s, 1970s, and 1980s. Nevertheless, there is still some development activity in the watershed. The fact that two of the founding members of LRN are involved in real estate development has led to a level of communication with the developer stakeholder community that is not often seen in grassroots environmental nonprofits. Steve Herbert (2012) provided an example of developer involvement:

> We have a development down on 19th Street, where the Doubletree hotel is, to build a huge apartment complex behind it. It will be approved, I think. It is a vested right. If it required rezoning it would never be approved, but because it was there the developer can build some more apartments. Their stormwater plant was piping it directly into Linkhorn Bay [part of the Lynnhaven River]. The reason they can get away with it is because the state had passed a stormwater credit option. It was the developer's idea, but for us it brought home that the problem is that they planned to go to James City County and buy wetland credits up there and then pump their discharge directly into Linkhorn Bay. It is perfectly legal under the new state legislation. We

> called the developer and let LRN know about it. This is an example of the influence of LRN. They found out who the developers were and called and talked to them about it. The developers came back in and said they were going to do stormwater management on site. Ten years ago in the city, it wouldn't have happened.

LRN also viewed landscapers and lawn-maintenance companies as stakeholders in the cleanup of the river. Forget noted that people are addicted to grass and their lawns. Getting people to see that they can enjoy their yards without perfect grass has been a challenge. "Fertilizer is a huge source of nitrogen and phosphorus, especially in a residential community like much of the Lynnhaven watershed is" (Forget 2012a). LRN has done some training with landscaping companies and lawn-care companies to try to tackle the problem from that perspective and has received good cooperation from the companies. "I think we are starting to make some headway on the idea that maybe having a perfect green lawn is not the best idea" (Forget 2012a).

Conveners

LRN's conveners could be identified as the three founders, Fine, Lester, and Stanton, along with the Virginia Beach city manager Spore and the first executive director, Frankenfield. LRN began as a citizen-oriented organization that excelled at leveraging resources and generating grassroots participation through its committee structure. The founders recognized that the mission of cleaning up the Lynnhaven River could not be independently accomplished by a single organization. There was a need for a collaborative partnership between the nonprofit, public, and private sectors in order for LRN to accomplish its mission. The City of Virginia Beach was not only a stakeholder, but it was critically important that it also serve as a convener. Very early in the process of establishing LRN, the founders met with Spore and convinced him that a partnership between the two organizations was necessary to accomplish the mission.

The founders of LRN knew who the key stakeholders were and they had informal ties to many of the citizens and businesses that would need to be involved in the effort. Interviews with the founders and city staff suggested that it would have been very difficult for the city, acting on its own, to establish these same informal relationships. The founders of LRN and its first executive director, Frankenfield, cultivated the collaborative process by incorporating city personnel and citizens into LRN's committee structure. Over time, based on their expertise and local connections with citizens, LRN came to be viewed as the legitimate convener of the collaborative arrangement, evolving into a "mixed group" comprising a diversity of stakeholders, most notably, the city.

The role of the three founders as conveners cannot be overemphasized. It was their passion and commitment to the goal of a clean river that brought the stakeholders together to work on solutions. Bringing city staff in as co-conveners was a brilliant strategy that immediately paid dividends. LRN could not be successful without the collaborative partnership with the city—"buy-in" on the part of the city was essential. It is to his great credit that Spore recognized that a mission of this magnitude required that the city government be involved from the beginning. He made it clear that the city was committed to the LRN mission by allowing a city employee to serve as the first executive director of the organization on a part-time basis. The city's role in the partnership would involve spearheading projects and providing support that did not directly involve citizen education or marketing efforts.

Resources

The collaboration of several partners, including the City of Virginia Beach, the state, and several nonprofit organizations, led to the creation of eight intertidal oyster reefs in the Lynnhaven River that covered six acres. The success of these reefs motivated USACE to make the Lynnhaven River a focus of a larger oyster-restoration program (Sorabella 2011, 14). In 2008, the USACE constructed more than thirty acres of subtidal oyster reef habitat in the Lynnhaven River and another twenty-two acres of oyster reef habitat the following year. According to Sorabella (2011, 14), "[t]he Army Corps hopes to conduct one final project in the Lynnhaven, the construction of more than 20 additional oyster habitat acres. Then it will be up to Lynnhaven River NOW and the community to build 'oyster shorelines' along waterfront property to reach the magical 100-acre goal." Partnerships with the City of Virginia Beach, state agencies, and USACE are perhaps the most important resources available to LRN.

From a financial point of view, the organization began with contributions raised by the three founding members. As described earlier in this chapter, the founders were able to obtain commitments from more than thirty people to donate $1,000 per year for three years in order to give the new organization a stable initial base of funding. A second initiative that provided early funding for LRN was a contractual arrangement entered into with the City of Virginia Beach to provide public education. This initiative was actually proposed by Charlie Meyer, a deputy city manager at the time, who saw the contract as a legitimate way to financially support LRN. As one interviewee put it, "environmental cleanup is not a problem in need of a solution, but rather a solution in need of funding" (Bernick 2012).

During its ten-year existence, LRN has continued to grow in financial strength, although in 2010 the organization did run a small deficit. The latest available Form 990 that the organization filed with the Internal Revenue Service in 2010 indicates that total revenue for LRN was $226,298 against total expenses of $230,767 for a net deficit of $4,469.

Salaries and benefits accounted for $145,307, and other expenses equaled $85,460. Net assets for LRN totaled $150,966 in 2010. The organization now has four full-time and two part-time employees and nine voting members on its governing board. There are approximately six thousand dues-paying members, and about one hundred people volunteer their time to LRN on a regular basis.

LRN has a well-designed website that provides a great deal of information about the organization and also encourages people to make a donation or to become a member—or renew their membership. There is an e-newsletter sent out twice a month and a sophisticated print newsletter sent out four times a year to interested parties. Forget noted that the investment in the newsletter seems to be paying off, "because then they get our annual giving letter—those names and addresses are very valuable. We get a pretty high percentage return from those annual giving letters—so the next time they get the mailing, maybe they become a member" (Forget, 2012b). Forget (2012c) also noted that "what has been really successful is the kind of cooperation we have gotten from garden clubs and groups like that. A lot of garden clubs have come in and gotten enough copies for their members. Some of the civic leagues have helped out, too" (Forget 2012c). Despite the fact that LRN has been in existence for a decade, Forget sees no signs of "donor fatigue." She noted that the Pearl Home Program has over two hundred homes and continues to grow and that the Pearl School program had thirty-five schools participating in 2011 and also continues to grow (Forget 2012c).

LRN has contracts with the Virginia Beach Department of Public Works to do stormwater education and awareness programs and also receives some funds from the Oyster Heritage Fund. Dave Iwans & Associates has been an important resource. From the beginning of the organization, Dave Iwans & Associates has donated its time working with the public relations and marketing committee. LRN pays direct costs for printed material, but newsletters, brochures, and other promotional materials are laid out and "made to look beautiful" by Dave Iwans & Associates. The Runnymede Corporation, of which Fine is president, has donated office space for LRN operations since the very beginning of the organization. The Dollar Tree Corporation has become an important partner for LRN. For the past five years Dollar Tree has funded the wetlands and classroom programs, which has allowed LRN to expand these programs. Forget (2012a) noted that:

> The nice thing about working with a local company like Dollar Tree is that they will fund it as long as they are happy with the program, whereas federal and state grants at the most would be a three-year grant. It will be funded to get it started, but never funded as an ongoing thing. Federal and state grants are great for running pilot projects and trying something out. The shell-collection program we do was originally a three-month pilot over a summer season to see how it

> would work with the restaurants and that was a state grant that funded that project. We were able to get ongoing funding for that so we could continue it. You have to have a combination of funding. You can't rely only on state and federal grants. They are great for trying something new and for money for restoration work and a project that is a one-time effort. For ongoing community and education you have to have other sources of funding.

An important resource for any grassroots environmental group, as noted in our framework, is access to expertise. The committee structure of LRN has allowed the organization to attract people with specific areas of expertise to work with and serve on the individual committees. The oyster restoration committee, for example, has representatives from the USACE, the CBF, and the Virginia Institute for Marine Sciences who are all experts in the field. In addition, the Virginia Conservation Network serves as a valuable source of information and expertise for smaller environmental groups like LRN. The Choose Clean Water Coalition is doing much the same thing at the federal level and puts out a book every year that groups like LRN find to be a very valuable resource.

As a final comment on resources, Forget noted that you have to grow the organization carefully. "We can hire ten people and keep them busy, but having the financial base to support them has to be there. It has to be a collaboration of volunteers and paid staff. It has to be a balance. You can't have so many volunteers that you can't manage them. Every committee has always had a staff person on it. We have always had that connection" (Forget 2012d).

Rules

LRN has a formal set of bylaws that underwent a major revision in February 2012. When the organization began a decade ago, it started with a rudimentary set of bylaws. The Articles of Incorporation and the bylaws were revised to better reflect what the organization is doing now. As Forget (2012d) noted, "You have to put those pieces in place. First of all, you have to do things legally. You want to make sure you get your 501(c)(3) and your tax returns and you are meeting all of those legal obligations."

According to the bylaws, LRN is organized exclusively for charitable and educational purposes. The property, affairs, and business of the organization is managed by a board of directors. The board oversees the formation of plans and policies affecting the organization, reviews and approves all programs and promotions submitted to LRN for sponsorship, and establishes short- and long-term strategic plans for the organization. The biggest group is the executive program committee, which is the primary working committee of the organization. This committee is composed of the chairs of the other committees and representatives from several of the city departments, as well as a number of members of the board of directors. This committee meets five times a year, every other month, but no meeting in December. The board of directors meets in the opposing months.

At the executive program committee meeting in July 2012 there were nineteen attendees. The meetings are fairly formal in nature, and minutes are kept by LRN staff. The meeting in July began with a welcome from Bob Wells, who is the chair of the executive program committee. Minutes from the previous meeting were approved, and Forget then conducted the rest of the meeting, making sure that the discussions and reports stayed on point. The meetings are an opportunity for all participants to discuss the important issues that are currently priorities for the organization. Typically reports are made by each of the committee chairs as well as the Department of Public Utilities and the Department of Public Works. There are no formal rules governing LRN's relationship with other organizations, but Forget does report on the activities of the city's Clean Water Task Force and Green Ribbon committee.

The executive director is responsible for hiring and firing the staff and proposes the budget and also serves on the board of directors. The board of directors approves the budget and is responsible for hiring the executive director. Other duties of the executive director include the general management and direction of the organization subject only to the ultimate authority of the board of directors. The executive director also presides over meetings of the board and acts as secretary of the board (Lynnhaven River NOW 2012b). The twelve-page set of bylaws of LRN is a comprehensive set of guidelines and operating rules that succinctly lays out the path forward for the organization.

Governance Structures

Institutional structure, as noted in chapter 2, is very important in a collaborative setting. In the case of LRN, almost from the very beginning, the organization set up a committee system to carry out its mission. Each of the committees focuses on one aspect of the work of the organization and establishes its own meeting schedule and initiates its own project ideas. One full-time staff member serves on each of the committees (Lynnhaven River NOW 2012b). The committee structure has evolved over time, and there are now a total of nine committees. One of the nine committees is the executive program committee, which provides oversight and guidance to all aspects of the work of LRN. This committee is composed of the executive director, the full-time staff, and the chairs of each of the other committees. The eight operational committees are: the clean boating and clean marinas committee; the education committee; the landscape practices committee; the oyster restoration committee; the public relations and marketing committee; the public policy committee; the stewardship and access committee; and the wetlands committee.

Commenting on LRN's governance structure, Christy Everett, the executive director of the CBF's office in Hampton Roads, stated that:

> What I think has worked incredibly well with LRN . . . is an unbelievable model and if I were to create my own watershed organization I would do this first. They have about nine committees and those nine committees have maybe nine members each and a chair. Through this grassroots transfer you don't need those four staff members to do the work of ninety people. You have those ninety people in their neighborhood thinking creatively, using their resources and reaching out to people to garner support. Otherwise, it is just a footprint of people. (Everett 2012)

The organization chart for LRN is presented in figure 5.2. The mission and primary goals of each committee are then discussed in the following sections, and the activities of some of the committees will be highlighted in more depth.

The Clean Boating and Clean Marinas Committee

Like many of the committees, part of the mission of this committee is education. The basic mission is to educate the public, boaters, and marina

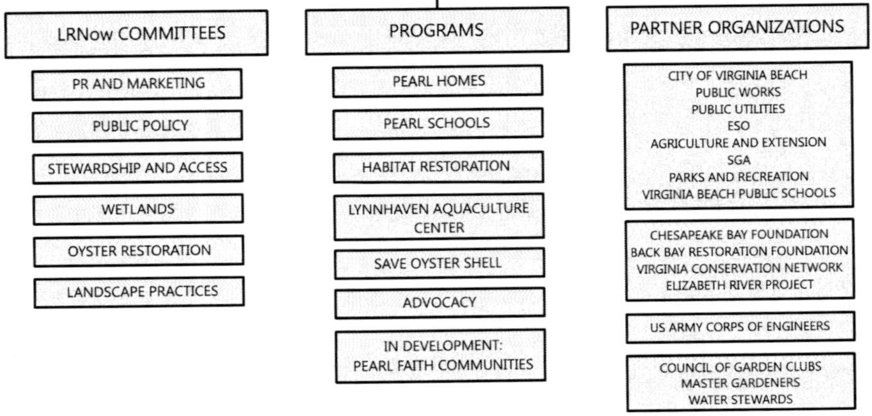

Figure 5.2. Organization Chart—Lynnhaven River NOW

owners and operators about responsible boating practices and compliance with the No Discharge Zone that was established on the Lynnhaven River. The committee also encourages and supports marinas on the river in attaining the Clean Marina certification. Through the efforts and assistance of the committee, there are now four "Clean Marinas" on the Lynnhaven River.[2] The committee was also responsible for the design, printing, and distribution of a No Discharge Zone brochure as well as obtaining the No Discharge Zone designation for the Lynnhaven River.

The Education Committee

The primary mission of this committee is to develop educational programs, materials, and opportunities concerning the Lynnhaven River's history, environmental issues, and efforts to improve the health and cleanliness of the river (Lynnhaven River NOW 2012a). This has been a very active committee that has initiated a number of important projects. Sorabella, the second executive director, working through LRN and her own small nonprofit organization, Oyster Reef Keepers, developed a school curriculum to teach children how to raise oysters and enlisted a large number of teachers to participate in the various school programs. The committee has sponsored summer teacher workshops and developed a school-recognition program, the Pearl School Program, to spotlight schools that made an extra effort to infuse aquatic education in their school programs. In addition, many of the school groups participate in oyster and wetlands grass restoration efforts on the Lynnhaven River.

The Landscape Practices Committee

This committee works to advise and educate the general public on sustainable landscape practices in order to improve water quality in the watershed. Workshops are held several times a year to teach citizens how to build rain barrels that can help reduce stormwater runoff from individual properties. Other workshops are also held to provide know-how to citizens on building rain gardens and buffer gardens. The committee is also actively involved in providing landscape-management information to citizens, gardening groups, and landscape professionals.

The Oyster Restoration Committee

This committee has a broad mission to bring together scientists, government agencies, nonprofit organizations, the public, and industry to generate ideas and develop programs to restore the native oyster population in the Lynnhaven River. Sorabella and Richard Serpe are the cochairs of the oyster restoration committee. Serpe noted that the oyster committee "has become a clearinghouse for information regarding oysters and oyster restoration for multiple agencies and institutions" (Serpe

2012). One of the big ongoing initiatives of the committee is an oyster "spat on shell" facility to be set up under the Great Neck Bridge on the Lynnhaven River.[3] This facility is being developed in conjunction with the city and will also serve as a demonstration facility that schools can visit for educational purposes.

The committee has been actively involved in other activities such as: oyster-reef construction on the river; promoting community oyster gardening; and the "Save the Oyster Shell" for the Lynnhaven program. Over the years the committee meetings have been attended on a regular basis by representatives of the USACE, VIMS, the Virginia Marine Resources Commission, and the City of Virginia Beach. As Serpe noted, the committee works hard to stay abreast of the latest science in order to promote oyster restoration. Different techniques for restoration have been tried on a small scale to see what works and what doesn't (Serpe 2012). From a collaborative standpoint, Serpe pointed out that "Lynnhaven River NOW has been very adept at reading the public and what can reasonably be done in terms of having the public behind you and not have the organization become an adversary in the process" (Serpe 2012).

The Public Relations and Marketing Committee

The mission of this committee involves creating and encouraging community involvement in improving the water quality of the Lynnhaven River. The committee helps with the printed material produced by the organization and works on television ads and other media efforts. The committee has also been responsible for merchandise such as hats and t-shirts and had a major role in the production of a beautiful "coffee table" book entitled *The Lynnhaven: Restoring a Legend,* published in 2011. The committee is chaired by Mary Reid Barrow, who has served in that capacity since the beginning of LRN. She is a naturalist who has written for the *Virginian-Pilot* newspaper for many years. The public relations and marketing committee is small in number, but does have several representatives from the city government. In November 2012, the committee produced a three-minute video that won a $5,800 prize from Whole Foods in a competition it sponsored for community organizations (Barrow 2012). The television ads produced by the committee have been memorable. They were designed to grab the attention of the viewer and at the same time to reinforce river-friendly behaviors that citizens could emulate.

The Public Policy Committee

The mission of the public policy committee is to promote dialogue between the organization and elected and appointed officials. As Mary Picardi (2012), the committee chair, put it, "The goal is to educate ourselves and our lawmakers about issues of importance to the health of the

River." The current public policy committee was organized in early 2009. Picardi was asked by Forget if she would be willing to take on the job of committee chair and she accepted. Picardi holds a master's degree in environmental medicine and is a retired physician. She lives in Virginia Beach and began attending meetings of LRN about five years ago. The public policy committee is a small committee and sometimes has a little difficulty attracting members. Picardi noted that it was hard to find people who were interested in talking with legislators in Richmond, the state capital. "As far as I am concerned," she observed, "the environment shouldn't be a partisan topic. It is, but it shouldn't be. People are kind of afraid of it and really don't want to go to Richmond and talk to people" (Picardi 2012). Forget (2012d) noted that "public policy was the biggest stretch for us in terms of how it fits into our core mission as an organization." LRN produces an annual state of the river report that allows the organization to evaluate where it is in reaching its goals.

The committee, like LRN as a whole, tends to attract passionate people. Picardi doesn't live on the Lynnhaven River, and when she was asked about her motivation for joining the organization she stated that:

> I am connected to the earth. This is my environment and earth. This is my little piece of it that I can do something about. Something that Mother Theresa said, "We may not be able to do great things on this earth but we can do small things with great love." That is what grassroots is. It is small things with great love and passion. (Picardi 2012)

Despite the small size of the committee, it has a number of ongoing activities and accomplishments. The committee sponsors a forum for candidates running for Virginia Beach City Council, and during the last Fall Festival the committee presented information to support a ban on uranium mining in the western part of Virginia as well as supporting the ban on the proposed Surrey coal plant, both of which would have a negative impact on the Lynnhaven River.

The Stewardship and Access Committee

This committee has a straightforward mission of promoting public access, enjoyment, and stewardship of the Lynnhaven River. The committee has a long list of goals that includes: support for open space initiatives; "Adopt-a-Spots" in neighborhoods, business districts, and communities in the Lynnhaven River watershed; educational signage at all access sites, where appropriate; and acquisition of a fleet of boats to use for leaders and volunteers to aid in river cleanup efforts.

Lillie Gilbert and Whit Peace are the co-chairs of the stewardship and access committee, and they have worked together on river projects for many years. Peace (2011, 23) remembers that:

> When it was my turn to give back, I was serving as Scoutmaster of Troop 419 while also working with the youth at Virginia Beach Friends School and as the Friends Meeting's Community Service Coordinator. I met this remarkable lady, Lillie Gilbert, owner of Wild River Outfitters [a local retail store]. It wasn't long before I was teaming with Lillie to get Scouts, Quakers, and Friends School students involved in cleanups at the Wild River's Adopt-a-Spot location on West Neck Creek.

Peace is the person who takes the lead on Lynnhaven River cleanup projects, which is a big part of the work of the committee. In an interview with Peace (2012) he noted that:

> When I first got started I thought we could keep doing what we were doing. What you have to do is take your energy and love for the earth back out there and spread it. I am trying to get more people involved. . . . The thing about getting young people involved, they go home and tell their mom and dad not to litter anymore.

Many of the committee's river cleanup efforts involve parents and children working together to remove trash and litter from the banks of the river. By all reports, the committee has been very successful in its efforts, but it is a never-ending job that requires continuing dedication and focus.

The Wetlands Committee

The mission of this committee is to conserve, restore, and create open water marsh, small marsh parcels along the shoreline, and also living shoreline marshes along the Lynnhaven River. Kevin DuBois and Joe Scalf are the current co-chairs of the committee. DuBois works for the City of Norfolk, but lives in Virginia Beach in the Lynnhaven River watershed. He is an expert in wetlands restoration and does that type of work for Norfolk. Scalf is a staff member at LRN and was hired as a restoration specialist; like DuBois, he has many years of experience in wetlands restoration efforts. The committee is a relatively new one for LRN, but has already been very active with about twelve to fifteen regular members.

When the Wetlands Committee was formed it was given two very specific goals, according to DuBois. The first was the creation of educational workshops for citizens. DuBois had already developed and taught similar workshops for the City of Norfolk and had been running them for a few years. The workshops involved both presentations and actual exercises in the field. The workshop presentations covered such topics as: the function and values of the wetlands; the regulations that govern wetlands and how they are implemented in the city; and living shorelines, which involves the use of native vegetation to stabilize the shoreline. DuBois (2012) noted that "I am passionate about wetlands so if I could spread that to the City of Virginia Beach, I was happy to do that." The wetlands workshops are designed to run every year. As DuBois further noted, "It's about empowering people to go out and influence other people. . . . All you have to do is get one person on the block and then neighbors talk. The key is to find out who are the influence peddlers in your community and how do you get to those folks. Who is it that everyone is talking to? You want to get to that person" (DuBois 2012).

The second goal was to try to convince homeowners on the Lynnhaven River to use native wetlands vegetation to stabilize their property instead of using hard structures like bulkheads. The committee was involved in the construction of the first public demonstration site for native wetlands plants at First Landing State Park on the Lynnhaven River. The plan is to create more of these sites over time. The committee also helped create a six-page wetlands brochure that talks about living shorelines and other issues important to wetlands restoration. As with many other volunteers working with LRN, DuBois gives a great deal of credit for the organization's success to the cooperative partnership between the city departments and LRN. In his words,

> [T]he executive committee works hand in hand with all the city departments. They have great support from the mayor and city manager. I think part of that is because of the founders of Lynnhaven River NOW. They come from the business community and they have those connections. They have figured out a really solid working relationship and that makes them very successful. (DuBois 2012)

Collaboration and the LRN: A Summary

The founders of LRN were determined that the organization they gave birth to would be one with a culture of collaboration and partnership. They recognized that their small organization could not tackle the problems associated with the cleanup of the river in isolation. No individuals or organizations were excluded from involvement in proposing or implementing programs that would benefit the effort of river cleanup. LRN offered the means for concerned citizens to become involved in finding solutions to the polluted state of the river and also created a path forward for the city to play a major role without appearing to be "heavy-handed" and forcing solutions on citizens. Collaboration requires respect for the opinions of others and a strong dose of civility, which is precisely what the founders of LRN brought to the table. The culture of collaboration is still very much the central philosophy of LRN after a decade of operation.

COLLABORATION OUTPUTS AND WATERSHED OUTCOMES

LRN can point to a number of successful outputs and outcomes from its activities over the past ten years. It is clear, after interviewing many people associated with LRN, that the organization does not claim all the credit for the improved state of the river and the return of oyster harvesting in the Lynnhaven. However, the collaborative process it set in motion has resulted in significant gains in the health of the river.

Outputs

Many of the outputs of LRN have already been discussed in this chapter. These outputs, as well as several others that have not yet been described, are presented in the following sections.

Plans, Agreements, and Partnerships

The partnership with the City of Virginia Beach was, and still is, vital to the success of LRN. From the very beginning the city supported LRN through contracts. The major contract, which is still ongoing, is administered by Public Works and provides funding from the Storm Water Utility Fund. The revenue for this fund comes from fees paid by property owners for stormwater services. A portion of this fund is provided to LRN through a contract to provide community and stormwater education. Forget noted that this contract involves almost everything that LRN does, such as Pearl Homes, workshops, rain barrels, and the "Scoop the Poop" program (Forget 2012b). A second, much smaller contract, funds the "Save our Shell" program and is administered through the Planning Department with money collected as part of the Oyster Heritage Fund.

Another aspect of the partnership with the city was the formation of the Green Ribbon committee. This committee was primarily the result of an initiative by the founders of LRN. According to Frankenfield (2012), the LRN founders wanted a committee to develop recommendations for action. A proposal to establish the Green Ribbon committee was taken to the Virginia Beach City Council and approved. The purpose of the committee was to review and make recommendations for improving the city's codes, ordinances, and practices with an eye to improving water quality. The committee, composed of city staff and citizens, met for more than a year to develop recommendations. In a report to the City Council, the committee made fifty-six recommendations; some were specific code and ordinance changes and some were training recommendations (Forget 2012a). Approximately forty-five of the recommendations were implemented, including a directive that all new buildings ten thousand square feet and larger in Virginia Beach had to meet lead certification requirements. Another example of the changes coming out of the Green Ribbon committee was a "Don't Feed the Wildlife" ordinance that was enacted to try to reduce the number of geese living in the city. The work of the committee also led to the development of a comprehensive stormwater management plan for the city which has resulted in the construction of improvements for stormwater one sub watershed at a time.

There is another important aspect of the partnership with the city government. In the words of Forget (2012a):

> We have a group called the Clean Waters Task force. It is a city committee that meets once a month and has for about seven or eight years now. It was originally the Lynnhaven River Task Force. The City Council passed a resolution in 2004 that set cleaning up the Lynnhaven River as a top priority for the city. Shortly after that the Lynnhaven River Task Force started. Our organization [LRN] was centrally involved in that task force from the beginning. We also included many other city departments. It evolved into the Clean Waters Task Force that still meets once a month.

The current mission of the Clean Waters Task Force has several elements. According to its mission statement the purpose of the task force is to:

- Provide an informal forum for information exchange on activities related to the city's watersheds;
- Help ensure better city agency coordination and cooperation on watershed-related activities and issues;
- Identify areas of concern that may need to be addressed by the city related to the city's watersheds;
- Foster a coordinated city response to other stakeholders working on watershed-related activities and issues;
- Support the efforts of individual participants in helping accomplish water-quality protection and enhancement;
- Help facilitate and implement the recommendations of the Green Ribbon committee, ensuring a measure of responsibility and accountability to the Green Ribbon committee's process;
- Designate use of Chesapeake Bay Preservation Area Board and Wetlands Board Civil Penalties Funds for restoration projects (Clean Waters Task Force 2012).

The task force not only includes representatives from more than ten city departments, but it also has representatives from almost all the nonprofit environmental organizations concerned with water quality in the city's watersheds. Two staff members from LRN participate in the task force deliberations, and the task force has proven to be an effective mechanism for coordinating many city-wide activities related to water quality in the Lynnhaven River and the other city watersheds.

For the past several years LRN has had a contract with the Dollar Tree Corporation to fund the Wetlands in the Classroom program. The Dollar Tree Corporation was in the process of putting together a corporate giving program, and it contacted LRN with a desire to get involved in some form of environmental educational program. Grants from the federal and state governments were an early source of funding for LRN. Sorabella noted that she wrote a grant that funded her position as education director and also launched a citizen oyster-restoration initiative. Later, in her first year as executive director of LRN, approximately $100,000 in grant money was raised (Sorabella 2012). Writing grant proposals is an arduous and time-consuming process, and there are fewer grant funds now available to nonprofit organizations. LRN still receives some grant funding, but the amount is much reduced from previous years.

USACE has been an important partner with LRN and the two organizations have worked closely together. Forget noted that LRN has done a number of joint projects with the Corps involving oyster restoration work in the Lynnhaven River. Next to the city government, the formal relationship with the Corps is the most important one that LRN has (Forget 2012b).

Scientific Reports and Projects

Each year since 2005 the major report issued by LRN has been the State of the River Report. This report tracks progress in four areas related to the river. The first is pollution, which includes separate categories for bacteria, dissolved oxygen, nitrogen and phosphorus, and water clarity. The second area is pollution control, which includes clean boating, sanitary sewers, new funds for water quality, and stormwater treatment. The third area is habitat, which includes categories for oysters, wetlands, open space and public access, and underwater grass beds. Finally, the fourth area is awareness, which includes media attention, membership and involvement, educational programs, and school participation. These areas, taken as a whole, cover all the major activities and projects of LRN. Each of these separate categories or areas receives a grade based on an assessment by LRN staff at the end of each year. The overall grade for the river, as assessed by the staff, has been a "B" for the last few years (LRN 2011). This grade is an indication of the significant progress that has been made in cleaning up the Lynnhaven River since LRN was formed in 2002. The Annual Report serves as a useful tool for LRN in the development of future strategies and directions to pursue. It should be noted that LRN is not the only organization that has produced science-based reports concerning the Lynnhaven River. The USACE, for example, has studied the river basin in depth and issued reports on ecosystem restoration and oyster-restoration initiatives.

In terms of specific outputs, LRN has executed a number of separate projects or programs that are designed to meet the organization's mission objectives. These projects and programs include: the No Discharge Zone and Clean Marinas; the Pearl Homes program; the Pearl Schools program; the rain barrels and oyster gardener programs; river cleanup projects; shoreline management; the "Scoop the Poop" program; the Save our Shell program; the alternative substrate initiative; the use of native vegetation; and wetlands workshops.

There are now six marine pump-out facilities for boats located on the Lynnhaven River. Since becoming a No Discharge Zone in 2007, over 6,500 gallons of boat sewage have been pumped according to the 2011 State of the River Report. The clean boating and clean marinas committee continues to promote both the No Discharge Zone and assists marina owners in attaining the Clean Marina certification through the State Clean Marina program.

The Pearl Home program was launched in November 2011 and has enrolled more than 650 families in the program (Forget 2012c). To earn the Pearl Home designation families must assert that they are doing at least fifteen different things from a list of forty-three behaviors that will contribute to improved water quality in the river. Pearl Home families receive a yard flag that designates them as Pearl Homes. The use of rain barrels is one of the things that homeowners can do as part of the Pearl Homes program. Rain barrel workshops conducted by LRN have been very popular with citizens. In these workshops citizens learn how to build rain barrels and can take them home at the end of the workshop. Rain barrels can also be purchased, and more than six hundred have now been sold. Another Pearl Home activity involves oyster gardening. In 2011 over three hundred thousand small oysters were stocked through the school and citizen oyster gardening programs (Annual Report 2011). As a thank you to the oyster gardeners, LRN conducted kayak trips where the oyster gardeners were able to place their oysters on oyster reefs in the Lynnhaven River. Forget (2012b) noted that Columbia Sportswear gave LRN a grant so that kayak trips were free to the gardeners. In 2011, forty-five schools in Virginia Beach qualified for a Pearl School award. To be eligible for the Pearl School award, schools accumulate points for projects outlined in LRN's Pearl School flyer. Schools must accumulate a specified number of points for the first-year award and continue to earn the requisite number of points in each consecutive year to earn and retain the Pearl School designation. More than 150 teachers participated in twelve teacher-training sessions in the last year (Forget 2012c). In addition, there are two programs that directly involve students on the ground. They are the oyster gardening program and the growing wetlands in the class program, where students grow wetlands grasses and transplant the grasses in the spring to wetlands restoration areas.

River cleanups, as noted earlier in this chapter, continue to be an important activity for LRN. Cleanups rely on volunteers; it would not be possible to do this work without dedicated volunteers. The stewardship and access committee has worked hard to get families involved in these activities although they primarily rely on word of mouth to get people involved. Scheduled cleanups are posted on the LRN website and promoted in the quarterly newsletter (Peace 2012).

The work of the wetlands committee has resulted in the Living Shorelines program. This program has been enhanced through the hiring of a new LRN staff position, the restoration project coordinator, which has given LRN the capacity to work much more directly with homeowners to promote living shorelines. LRN has now begun to identify sites for several model living shorelines and will work with homeowners who are interested in building living shorelines on their properties (Forget 2012c). LRN has produced a Tidal Wetlands Restoration brochure that provides a very useful guide to the importance of living shorelines and suggests tips for how to create them as well as listing sources for wetland and native plants. In addition to the brochure, the committee conducts wetlands educational workshops for citizens. The committee was also responsible for constructing LRN's first public demonstration site for the concept of living shorelines with native wetlands vegetation.

It is difficult to demonstrate the specific impact of the "Scoop the Poop" program. However it was one of the early signature programs of LRN and continues to be heavily promoted. In our interview with Sorabella (2012) she assessed the program, noting that:

> Without the Scoop the Poop campaign we would not have addressed some of the pet contribution to the river's pollution. I think in the past people were thinking of picking up after their pet as a courtesy to somebody else that was walking the streets. I don't think people realized they should scoop the poop even from their own yard to get it out of the stormwater. I think it made a tremendous difference. The awareness that you have to get the waste even off of your property is critical.

The Save our Shell program, operated by LRN, collected approximately 1,200 bushels of oyster shells from restaurants, events, and the public in 2011. These oyster shells are useful in building oyster reefs, and plans are now under way for the construction of twenty acres of new sanctuary reefs (Annual Report 2011). LRN is also working with the city on an initiative to build an oyster spat on shell facility underneath the Great Neck Bridge on the Lynnhaven River (Serpe 2012). Another initiative related to the Save our Shell program was the search for alternative substrates for the construction of oyster reefs. Oyster-reef restoration requires a substrate that oyster spat can attach to, and there are not enough oyster shells available to build all the reefs. An alternative to oyster shells was scrap granite pieces left over from the production of countertops. Sorabella contacted one of the companies producing countertops, Rock Solid Surfaces, and asked for granite scraps. The company was more than willing to provide the scraps and could deliver two dump truck loads per week. The granite scraps should serve as an "in kind" match so that the USACE can proceed with oyster-reef restoration work and meet federal matching requirements. According to Sorabella, the granite will be useful even if the oysters don't settle as well on granite since it can still be used as a foundation for the reefs that could then be capped with oyster shells (Sorabella 2012).

Water-Quality Standards

The testing of water quality in the Lynnhaven River is conducted on a monthly basis by the State of Virginia Department of Health Shellfish Sanitation Division at thirty-six specific locations on the river. They also test after major rain events and hurricanes. The river is shut down for shellfish harvesting automatically after these events for a period of seven days. They will either extend the shutdown or lift it depending on test results. The agency evaluates the results of testing twice a year in November and March in order to draw new maps showing which areas of the river are open to shellfish harvesting. Oysters must be in clean waters for a specific number of days or they cannot be sold (Forget 2012d). LRN has not gotten involved specifically in the testing of the river's water quality since this function is already being competently performed by the state agency.

Education and Awareness

LRN has excelled in its mission of raising environmental awareness in the Lynnhaven watershed. Community education is at the core of LRN's work. It is a growing organization with more than six thousand members and volunteers at the end of its tenth year of existence. As noted in the 2011 Annual Report, more than eighteen thousand people were involved with LRN through a wide range of educational, outreach, and volunteer events and activities during the year (Annual Report 2011). The year 2011 saw the publication of LRN's first book, *The Lynnhaven, Restoring a Legend*, with close to five hundred copies distributed by year's end. Additionally, LRN has produced quarterly newsletters, brochures, and calendars. The public relations and marketing committee has produced television ads and other promotions to raise citizen awareness (Barrow 2012). Many newspaper articles have been written in the last few years about the activities of LRN, many of them written by Barrow, chair of the Public Relations and Marketing Committee.

The Oyster Roast, held every spring, is a major fundraising event for LRN and usually attracts five hundred participants to the event. In 2011 LRN held its first Lynnhaven River Art Show and its first fall festival. These events were very successful and well attended. The fall festival was held again in November 2012 at Virginia Beach Middle School and featured activities for children, live music, displays for adults, and talks on a variety of topics related to LRN programs. There were a number of other events held during the year to promote citizen awareness, including the "Fall on the Lynnhaven Bike Tour" and candidate forums cosponsored by LRN. Other outreach channels include the organization's sophisticated website and presence on social media sites such as Facebook and Twitter. Additionally, through its public policy committee, LRN maintains contacts with state legislators in Richmond and lobbies for legislation favorable to the health of the river and the environment in general.

Outcomes

The most important outcome that can be attributed, at least in large part, to the activities and programs of LRN is the reduction of bacteria in the river. In LRN's first State of the River Report in 2005, it was reported that only 1 percent of the Lynnhaven River was open to shellfish harvesting due primarily to bacterial contamination. In the most recent State of the River Report for 2011, it was reported that over 42 percent of the river is now open to shellfish harvesting (Annual Report 2011). This is a tremendous accomplishment that has astounded the founders of LRN (Fine 2012). The vision of the founders, which was encapsulated in the original name of the organization, Lynnhaven River 2007, was that oysters from the river would be suitable for harvesting and eating by the year 2007, which happened to be the four hundredth anniversary of the first landing by English settlers in 1607. The goal was met, and in 2007 enough of the river was open to shellfish harvesting that people attending the 2007 Oyster Roast were able to consume oysters from the Lynnhaven River.

A second important outcome, very much related to the first, was the return of commercial watermen to the Lynnhaven River. The absence of commercial shellfish harvesting in the river could be measured in decades. With more than 42 percent of the river now open to harvesting, there are several companies growing and harvesting oysters on the river on a scale large enough to support their operations. Watermen attribute their return to the river to the work of LRN in partnership with the City of Virginia Beach (Chalmers 2012).

The first two outcomes would not have been possible without actual changes in the behavior of the residents of the Lynnhaven River watershed as well as citizen support for stormwater infrastructure improvements made by the city. The many programs and educational activities of LRN have served to build citizen awareness of the impact of their behaviors, such as landscape practices and cleaning up after pets, on the health of the river. The need for reinforcing and increasing citizen awareness will continue to be a major part of LRN's mission, but the accomplishments to date have been impressive.

SOCIAL CAPITAL

This case study of LRN clearly illustrates that citizens in a community can come together to accomplish goals that seemed impossible to accomplish in such a short period of time. There is still much work to be done before the Lynnhaven River can be considered pristine; however, the accomplishments to date have astounded even those citizens working directly in the effort to clean up the river. Because the Lynnhaven watershed is almost entirely residential, the cooperation of many thousands of residents in changing their behavior was, and is, absolutely essential to make any progress toward a clean river. The willingness of many citizens actively to participate in the mission of LRN is evidence that a significant level of social capital existed in the City of Virginia Beach prior to the formation of LRN. The founders of LRN recognized that the success of the newly formed organization would be dependent on committed volunteers and citizens of the community. In an interview with Bob Stanton (2012), one of the founders, he noted that:

> I think when you finish your book it is going to be a marvelous testimonial to people working together. This is what can happen when people are working together and get their eyes on the same ball. Egos do not get in the way. People get to the point where they do not care who gets the credit; they just want to get it done. I don't remember any kind of grandstanding.

The founders reached out to many individuals and organizations, including the city government, to get them involved in the mission of cleaning up the river, and this went a long way to provide an initial base of social capital to build on. All three founders have very strong ties to the community and they are well respected and trusted, which has been a major asset in fostering and enhancing social capital. There is a strong foundation of civic leagues and other citizen organizations in Virginia Beach that contributed to the initial level of social capital in the community. When LRN was formed there was a strong community response in terms of contributing both volunteer time and money dedicated to the cleanup of the river. As Forget noted, "Not being able to eat oysters from the river any longer was a big deal in the community" (Forget 2012a). Many groups and organizations partnered with LRN, including school groups, to achieve the goal of a cleaner river, which is another indication of the level of social capital in the community.

As with the other case studies presented in this book, the participants in the LRN interviews were asked specifically about the contribution made by the organization to the level of social capital in the community. Participant responses include the following remarks:

> The early years when I was here I would go to statewide meetings and we were viewed as this giant sprawling "don't give a damn about the

environment," suburban cancer that ought to be obliterated from the face of the earth. I remember in the newspaper, people from Chesapeake said, "We don't want to be like Virginia Beach. We want to learn from that mess. We don't want to repeat those mistakes." You don't hear that anymore. If you go around to the state organizations, Virginia Beach is now perceived totally different. It wasn't all LRN, but it was part of that. It was part of that movement that started it out. We are working on our sustainability plan and we are working on our carbon footprint and alternative energy. We are working on all kinds of issues that we would have not been able to touch politically fifteen years ago. There are so many more people involved in it. It is a whole cluster of these kinds of things that are working and getting aligned with some of these environmental issues. LRN really kind of jump-started that whole community engagement process. To me, that is social capital. (Spore 2012)

I think the transparency of LRN has gone a long way to build trust between the city government and LRN and I think it has also helped to move the city forward. (Bernick 2012)

LRN is part of the social fabric of the city leadership, neighborhood leadership and they have roots into a lot of different organizations. My mother-in-law, for example, is determined to make her house a Pearl Home out on Bay Island. She is eighty-three years old and all into this stuff. She is following this and she knows she can call LRN and get information. She talks to her knitting club and church group about it. LRN is well known and very influential at many different levels. Yes, I think it is a phenomenon that they have very knowingly worked at to advance their goals. (S. Herbert 2012)

I think one of the smartest things that LRN has done is invest heavily in the relationship with kids. One of my great hopes is that we are going to get back to near historical pristine levels in the river based upon the role kids will play in the future when they are presented with decisions. It isn't to say that we haven't gotten an excellent response in terms of stewardship for the river from the existing generation. (Serpe 2012)

The river cleanups are the most amazing things in the world. The people who participate in those, I mean, it is filthy, dirty, horribly hard work and we do them year round every couple of months. I think to get people involved in something like that is pretty remarkable. (Barrow 2012)

If you belong to LRN, you can go to workshops, get involved in river cleanups, learn to grow your own oysters. I think most people like to be involved in what they think is a higher goal. They are concerned about their environment, they see the decline around them and want to make a difference and you have an organization that gives you the ability to do that—it's great. (Johnston 2012)

There is no doubt, as the comments above illustrate, that there is a widely shared belief that LRN has contributed to the social capital of the community. From the very beginning of LRN the organization was seen as a legitimate representative of environmental interests on the Lynnhaven River. The founders of LRN had a long history as influential community stewards with access to other community leaders. They were able to raise a substantial sum of money to support the initial operations of the organization. As Bob Stanton (2012) noted, "We went to the people who live on the river and asked them, 'Would you rather have a clean river or a dirty river in your backyard?' It was fairly easy to build a fairly substantial base of financial support." People were willing to donate money and volunteer time because the new organization and its founders were seen as both legitimate and trustworthy.

Trust is one of the other important elements of social capital. Forget (2012b) reflected on the ten-year history of LRN:

> My impression is that we have a high level of trust in the community, that people trust our organization to be reasonable and to be promoting the best practices and to be ethical in the way we do things. We have a high level of trust with the city and one of the ways you keep these partnerships going is to always remember that it's a two-way street. I don't show up at a meeting just when I want someone to do something for us. We try to be a really active partner with all of these other groups in the community and I think what we can do for them is as important as what they can do for us.

Forget's statement is not only an indication of trust levels, but also speaks to norms of reciprocity engendered by LRN. Perhaps one of the clearest examples of reciprocity involves the watermen on the Lynnhaven River. Unlike the case of the NRPA, the watermen have a good relationship with LRN. As Forget (2012) noted, "Our oystermen love us because they are only in business if the water quality stays good. So we have a very good relationship with them." Several of the watermen donate oysters for the LRN annual Oyster Roast event. In addition, one of the oyster companies donates a certain percentage of its profits to LRN. Other businesses that rely on the river, such as kayak rental businesses, have also made a similar commitment. Dave Iwans & Associates, the Dollar Tree Corporation, and scout troops have all donated expertise, money, or volunteer time to LRN.

LRN has cause to celebrate in that its original goal of having an oyster harvest by 2007 was met. In fact, Lynnhaven oysters were served at the 2007 Oyster Roast, and since that time even more of the river has opened to oyster harvesting. Efficacy, as an element of social capital, shows the ability of the organization to accomplish its goals. In this regard, LRN has been more than successful and continues its work to clean up more of the Lynnhaven River.

Commitment to Place

A commitment to the Lynnhaven River was a common theme of the interviews conducted for this case study. This finding is in line with Edward Weber's (2009) finding concerning commitment to place discussed in chapter 2. The three founding members of LRN all had deep connections to the area and to the river. Their commitment is perhaps best illustrated by the comments of two of the founders. Fine (2012) recalls growing up near the Lynnhaven River:

> I had enjoyed all of the pleasures of playing games in this most gorgeous setting and when we were hungry we would go down and take oysters out of the river and either eat them raw, or finding a crate and making a little fire and grilling them. That is pretty firmly etched in my mind.

Lester (2012), another of the founders, discussed his connection to the river:

> I came to Hampton Roads in 1969 and I was a junior officer in the Navy and I used to go to the Duck Inn (near the mouth of the Lynnhaven River) and they had these aluminum trays and would shuck the oysters right in front of you and it was just wonderful. In our adult lifetime, because of all these forces . . . the runoff because the city grew and all the stuff came off the yards and streets, the pets, the animals, the stupid geese, and the runoff from people's lawns. We didn't really know what it was. All we knew was that we couldn't eat the oysters so we said, "I wonder if there is something that we could do?"

Lester also commented that:

> It is not a river in New Jersey, that is my river and I want to be helpful. There is social capital being built because people are feeling good about what they are doing. I did something good today, I picked up after my pet. I think it probably makes people feel better about themselves. People love success. People love to be part of a winner. (Lester 2012)

As Dubois (2012) noted, "I live on the river so I am a stakeholder in that regard." His comment was echoed by almost all the interviewees and helps to explain the high level of motivation to clean up the river that is so evident in the willingness of committee members and others to participate in the mission of LRN. It is not just LRN members and volunteers who have a commitment to the watershed and Virginia Beach. As Sorabella, LRN's second executive director, observed:

> Whenever we would be at a civic league or someplace like that, maybe they would say they were working on an environmental initiative and maybe you could help us with that sort of thing. There really was a nice exchange that we were able to do. They were helping us and giving us a forum for educating and talking to people. We could help them do

> some of their environmental initiatives too. I feel like there was some community engagement we were able to get with that. (Sorabella 2012)

SOME FINAL THOUGHTS

There is no doubt that LRN has accomplished a great deal in its ten-year existence. A number of factors have contributed to its success. To begin with, the founders adopted an organizational philosophy of nonconfrontational cooperation. Under their leadership LRN became a voice of environmental moderation in the community and avoided "attacking" other stakeholders. In a sense the organization demonstrated peer "modeling" behavior rather than peer pressure in its dealings with the community at large.

Closely related to the philosophy of cooperation is the inclusive culture of LRN. The organization welcomes everyone who wants to help with its mission of a cleaner river. No stakeholder group is excluded, and this goes a long way to creating greater levels of social capital in the community. Inclusiveness is demonstrated by LRN's willingness to form partnerships with the city government, the school system, and private organizations. This culture of inclusion has contributed to the high level of legitimacy and trust that LRN enjoys in the community. LRN operates in a transparent fashion, and it is not perceived as having "hidden agendas." The organization is further helped by the level of respect in the community for its well-known founders/conveners. These men brought several important attributes to the organization including: a passion for the mission; personal attachment to the river; demonstrated ability as fundraisers; and strong connections to the business community in Virginia Beach. All nonprofit organizations need champions. LRN was particularly fortunate to have such dedicated and effective ones.

The importance of an understandable, straightforward mission has been illustrated in the case of LRN. Volunteers, members, and stakeholders were energized by the articulation of a clear mission objective. The original name of Lynnhaven River NOW was Lynnhaven River 2007, which was actually a statement of the goal to be able to harvest oysters from the Lynnhaven River by the year 2007. Having measurable goals is important in measuring the success of an organization; however, LRN was unusual in that its most important measurable goal was actually incorporated in the name of the original organization. Harvesting oysters from the Lynnhaven River by the year 2007 was not only a measurable goal, but also served as a highly symbolic goal that served to galvanize the efforts of volunteers and stakeholders. An important aspect of the mission of LRN was that the organization did not try to affix blame for the polluted state of the river in 2002. There was no one defining event that caused the pollution. To put it simply, the degradation of the river occurred over a long period of time, and there was no point in trying to single out a particular culprit for the level of pollution in the river.

Maintaining the enthusiasm of volunteers and members is critical to the success of any nonprofit organization. LRN has done a particularly good job of this with its annual Oyster Roast event. This event, where oysters from the Lynnhaven River are served, demonstrates to all involved that their efforts are actually paying dividends and the overall mission of the organization is being accomplished. As noted earlier in this chapter, the Oyster Roast also serves as a fundraising and "friend-raising" event for LRN. The importance of celebrating success is that success can lead to even higher levels of participation in the activities of the organization.

An important aspect of LRN's success has been its principle of organizing around a committee structure with each committee taking responsibility for one aspect of the organization's mission. The cleanup of the river had to be based on strategies that involved science in order to be effective and to gain the public's trust. From the beginning LRN welcomed and even recruited individuals with science-related backgrounds to serve on its committees. Projects provided an opportunity for people with expertise to contribute their time and knowledge to a goal. In fact the second executive director of LRN, Sorabella, had an advanced degree in oyster restoration from VIMS. The level of expertise available to the organization was important in lending credibility to LRN's programs and activities.

Early on LRN recognized the need to get the next generation involved in and concerned about the environment. The importance of the investment in the education of school children became a key aspect of LRN's mission. Reaching out to children through school programs not only helps to change the behavior of children, but also helps to change the current behavior of adults. River cleanup efforts encouraged the participation of children and their parents, which can lead to behaviors that involve environmental stewardship of the river. Education programs for children can help to convince them that the current state of the river is not its normal state and that they should help work toward the goal of a pristine river.

A final point about LRN is that, as an organization, it has been more than willing to share what it has learned with other nonprofit organizations working in other watersheds. Although these types of activities do not directly contribute to LRN's mission, the staff sees this as an important contribution and responsibility to be collaborative in region-wide efforts for a cleaner environment in Hampton Roads. Key lessons from the LRN case are presented in textbox 5.1.

Textbox 5.1

Lessons from Lynnhaven River NOW

- Fundraising is critical to the long-term viability of nonprofits, but fundraising should also be seen as an opportunity for "friend-raising."
- Inclusiveness is important. Grassroots organizations that seek out and include all stakeholders will be more successful.
- Form working relationships with local companies. They are likely to continue funding programs as long as they are happy with the programs.
- Nonprofit organizations have to grow carefully. The financial base must be in place to support the growth.
- Structure matters. The organization of nonprofits is most likely to be successful when a committee structure is created that reflects the organization's mission.
- Collaborative effort requires respect for the opinions of others and a strong dose of civility in the conduct of business.
- Operating in a transparent fashion without "hidden agendas" builds trust between partners in a collaborative undertaking.
- Focusing on blame for the current state of affairs accomplishes nothing and can alienate stakeholders.
- Recognize that grassroots organizations can do some things better than city governments (education and awareness) and vice versa (undertaking major projects and regulations).
- Grassroots organizations can act as "force multipliers" for local governments and allow them to move environmental projects forward that they could not accomplish on their own.
- Goals: grassroots organizations should focus on a symbolic goal that grabs the public's attention in order to get volunteers involved.
- Early success, no matter how small, generates enthusiasm and commitment. Go for the "low-hanging fruit" to cement the collaborative effort.
- Starting new programs and generating new ideas helps to eliminate "donor fatigue."
- Availability of intellectual capacity and expertise is likely to lead to greater success for the collaboration. Grassroots organizations must attract and retain experts who will volunteer their time.
- Education is an important part of the work of grassroots organizations. Programs in schools get commitment from the next generation. When children care, parents care.

- Personal attachment to place is important. People care "because it's in my backyard."
- Shared learning is important. Grassroots organizations should share what works and what doesn't with other groups in order to produce a wider impact.

NOTES

1. A No Discharge Zone is a designated body of water where the discharge of treated and untreated boat sewage is prohibited under federal law.

2. The Virginia Clean Marina Program is designed to promote the voluntary adoption of measures that prevent or reduce pollution from marinas and boatyards. Marina operators that adopt these measures are designated as "Virginia Clean Marinas." Marinas must meet 100 percent of the regulatory and legal requirements as well as a significant percentage of the recommended practices in order to receive the "Clean Marina" designation. A sample of the recommended practices includes: maintaining shoreline vegetation; maintaining oil-spill response equipment; having environmental controls at fuel pumps; having a well-maintained and accessible marine sewage pump-out system; and having a stormwater management pollution prevention plan. Complete information about the Virginia Clean Marina Program is available at www.virginiacleanmarina.com.

3. Oysters attach to each other and other hard structures, forming dense reefs that provide habitat for fish and other creatures. Adult oysters release eggs and sperm into the water, where they are fertilized. In less than twenty-four hours, the fertilized eggs develop into larvae. For the next two to three weeks, the free-swimming larvae grow and develop a foot to crawl and "explore" surfaces. When they find a suitable surface to settle on, they secrete a cement-like substance, which fixes the left valve into place. Attached juvenile oysters are called "spat." It is hoped that growing spat on shells and introducing the oysters to the river when they are larger will reduce the predation rate from cow nose rays that feed on juvenile oyster shells.

SIX

The Nansemond River Preservation Alliance

A Tale of Two Rivers

So we found a tale of two rivers—the lower half, which was reasonably clean . . . and the upper half, which was listed as impaired even for swimming.
—Nicholas Worth (2012)

On a stormy evening in mid-May 2012, about eighty-five people gathered in an old schoolhouse in the Eclipse section of Suffolk, Virginia, to attend a "River Talk" sponsored by the Nansemond River Preservation Alliance (NRPA). Attendees brought home-baked cookies, pies, and cakes; others had brought bottles of soft drinks or *hors d'oeuvres*, and many had arrived forty-five minutes early for the evening's program. Friends greeted friends, and it seemed as though everyone in the room knew one another. The speaker that evening was Walter Priest, a wetlands restoration expert from the National Oceanographic and Atmospheric Administration (NOAA). The crowd, most of whom were middle-aged, listened attentively as Priest spoke for about forty minutes on the topic of shoreline-restoration techniques; many in the crowd took notes. In the twenty-minute question-and-answer session that followed, the questions were thoughtful and specific—this was clearly an audience that had paid close attention. When Priest spoke of the importance of getting children involved with wetlands restoration at an early age, many in the audience nodded knowingly; a few clapped. When the program ended, the crowd gave Priest a standing ovation. Many in the crowd stayed behind to talk to Priest; others stayed to socialize with friends, or help clean the meeting space, a process that lasted another hour. When asked about the convivial, pleasant "vibe" in the room, one attendee remarked, "We're all friends and neighbors; it's like getting together at someone's house—but with a purpose."

The story of NRPA is similar in some ways to those of the Elizabeth River Project (ERP) and Lynnhaven River NOW (LRN). Like LRN, the people behind the creation of NRPA are residents with a love of the river, important connections in the community, and the wealth to support the early activities of the organization. Like ERP, NRPA receives little direct support from local government. The water-quality issues in the Nansemond watershed are similar to those in the Lynnhaven (nonpoint pollution from suburban runoff). All three organizations espouse collaboration as the key to the realization of their goals, and all rely on committees to develop and implement programs.[1] In spite of these important similarities, the story of NRPA, a relatively new organization, is different from our other two case studies,[2] and thus provides a contrasting case study through which we may examine the efficacy of collaboration in local grassroots organizations.

THE CONTEXT AND HISTORY OF NRPA

The Nansemond River lies in the western part of Hampton Roads, and flows from the southwest to the northeast where it joins the James River at the confluence of the James River, the Elizabeth River, and the Chesapeake Bay (see figure 6.1). Like the Lynnhaven River, the Nansemond is located within a single political jurisdiction (the City of Suffolk). The river is about twenty-three miles from its headwaters to its mouth, and varies from about one hundred feet wide in downtown Suffolk to nearly two miles wide at its mouth. Several small creeks and streams feed into the river, and what little fresh water feeds into the river near its headwaters flows from the nearby Dismal Swamp. The water at the upper end of the river is high in tannins and low in salinity.

The City of Suffolk, formerly Nansemond County, was created in 1974 with the consolidation of the cities of Suffolk, Holland, and Whaleyville, and the county. With the consolidation, the newly formed City of Suffolk encompassed some 430 square miles, making it the largest city in land area in Virginia (City of Suffolk 2012). Nansemond County had been a largely rural area, perhaps best known as the home of Planters Peanuts. The original city of Suffolk, located in the south central part of the county, was located at the boundary of the navigable portion of the river. Much of the rest of the county was farmland, and farms still dominate the area between downtown Suffolk and the state line to the south. The northern part of the county was also farmland, although suburban development began to invade the northern areas in the 1980s as developers sought cheap land to build bedroom communities for Norfolk.

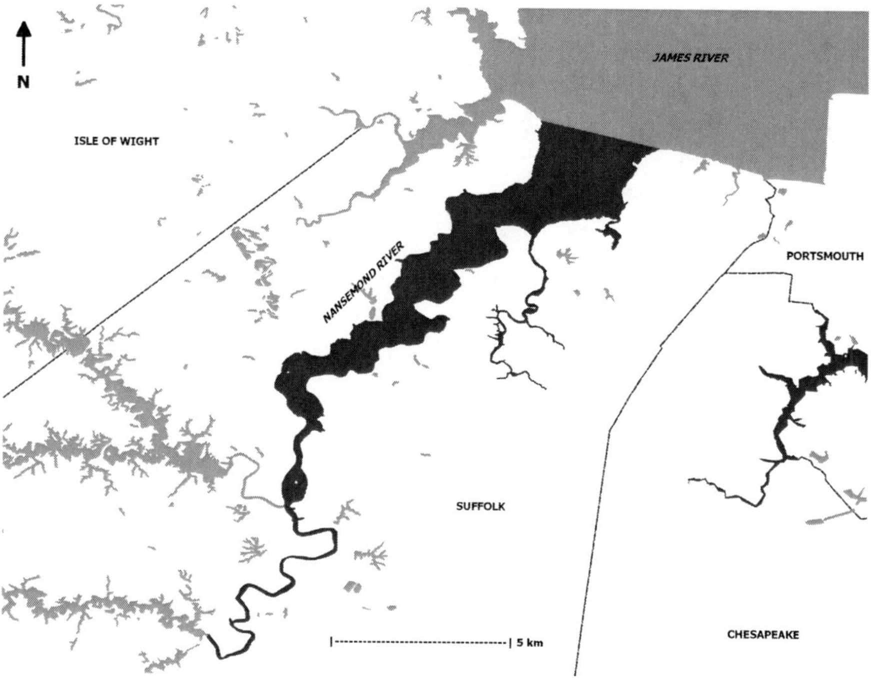

Figure 6.1. The Nansemond River Watershed

In the early twentieth century, three tributary streams near downtown Suffolk were dammed to provide freshwater reservoirs for the city of Portsmouth, just to the east of Suffolk. This action created three new bodies of water—Lake Cahoon, Lake Kilby, and Lake Mead (Nansemond River Preservation Alliance 2012a). This action had two important consequences. First, the river, which is largely tidal, no longer flushes adequately from its source to the mouth. The water near the headwaters tends to hold bacteria, nitrogen, and phosphorus, and the levels of these contaminants can be quite high. In fact, the river lost nearly 40 percent of its flow capacity when the dams were built (Taraski, Smith, and Wass 2012). Second, the areas lost as the tributary streams were dammed were largely wetlands, so the runoff into the river no longer must pass through a buffer area, but is dumped directly into the river. Coupled with the bulkheading[3] in developed areas and loss of wetlands due to development downstream, much of the runoff into the upper Nansemond is largely unfiltered. Significant rain events raise bacteria levels sharply and lower salinity in the river; the impact of these rain events is more pronounced due to the lack of natural flow in the existing river.

Pressures of development continued in the city through the 1990s and 2000s, as more developments were built along the US 17 corridor between Portsmouth to the east and the approaches to the James River Bridge to the west. This development also began to spread south toward downtown Suffolk; the population of the city has increased from 47,621 in 1980 (US Census Bureau 1982) to 84,585 in 2010 (US Census Bureau 2012).[4] Waterfront property is limited, and many of the farms along the river were sold for development. At the present time, there is only one area for public access to the river; a city park on Bennett's Creek (a small tributary to the river) offers two small boat ramps.

Political Culture

The local culture in Suffolk tends to be dominated by traditional farming interests, as well as those who wish to stem the growth of new residential development in the northern part of the city. Though the city is largely developed, many residents cling to a deep-seated desire to retain traditional conservative values. The dominant political culture closely resembles Daniel Elazar's (1984) traditionalistic culture, with perhaps a splash of individualistic culture thrown in as well. As is the case in all of the cities in Hampton Roads, Suffolk employs a council-manager form of government, but the city council is heavily dominated by traditionalistic interests. As a result, government services tend to be somewhat less expansive than those in the cities to the east.

As an example of this culture, in late 2011 the city of Suffolk instituted a policy designed to control the ability of city employees to interact with the public. In order for a member of the media, a citizen's group, or a researcher to talk to a city employee, permission must be sought from, and granted by, the city's Office of Media and Community Relations. Interviewees for this case study expressed some frustration at the added "red tape" this requirement created, and one suggested that "city government had lost its purpose—to serve citizens." This viewpoint is consistent with Elazar's description of traditionalistic government, marked by "a paternalistic and elitist conception of the commonwealth" (Elazar 1984, 118). Elazar goes on to describe this culture as one that is largely closed, self-perpetuating, and one in which "political leaders play conservative and custodial rather than initiatory roles unless they are pressed strongly from the outside" (Elazar 1984, 119).

The Physical Setting

The NRPA office is located in the Crittendon-Eclipse-Hobson (CEH) area in the northern part of the city, bordered on the east by the Nansemond River and on the west by Chuckatuck Creek. The small peninsula of land sits at the confluence of the Nansemond River, Chuckatuck Creek, and the James River. Just east of this peninsula the James River, the Elizabeth River, and the Chesapeake Bay come together in the body of water known as Hampton Roads, from which the region takes its name. The office itself is in an old schoolhouse building, which dates back to the early twentieth century. The building is owned and maintained by the Ruritans of CEH, a civic group that serves the local area. NRPA shares its space with the Suffolk River Heritage (SRH).[5] In addition, the room is used one day a week by a group of senior citizens who meet for several hours to socialize, play games, and engage in arts-and-crafts activities. The front of the room contains two desks, several small tables, file cabinets, and some bookshelves. Much of the rest of the room is taken up by several round plastic tables and plastic chairs, and shelves and cabinets full of art supplies. The building is not centrally heated, so there are two large natural gas heaters in the room. The rest of the building contains offices for the Ruritans, a storage area, a large kitchen, restroom facilities, and a large public meeting room with a stage at one end. There is a gravel parking lot outside, and a large grassy area out back where public functions are sometimes held. The area around the building consists of well-kept but modest homes, many of which were built in the 1950s and 1960s.

The link between NRPA and SRH is no coincidence. Karla Smith, one of the founding members of NRPA, had been an active member of SRH, and serves as its president. Smith's long-time interest in history, particularly the history of the region, was a significant factor in the formation of the historical group in 2005. Her connections in the community have played an important role in the growth and development of NRPA, as she acts as a bridge between the "come heres" (residents who have only recently moved to Suffolk) and the long-time residents. Smith and her husband have lived in Suffolk on Chuckatuck Creek for many years, and her husband, Jimmy, grew up in the area (Smith 2012b). Karla's engaging personality and openness, along with her strong sense of community spirit, make it easy for her to span the gulf between the native residents and the more recent residents, giving her a great deal of credibility among both groups.

Formation of the Organization

The NRPA is the youngest of our three case study groups, having been formed in late 2009. The genesis of NRPA was a conversation between Richard "Dick" Barry, a retired executive with Landmark Communications, and Marjorie Mayfield, the executive director of the Elizabeth River Project (ERP). Mayfield had been a reporter with the *Virginian-Pilot*, the local paper owned by Landmark Communications; when she was first starting ERP, she approached Landmark to become a sponsor of ERP:

> So she said, "Gosh, I know you so therefore you should give us some money." And we did. She was just very impressive the way she started

> [ERP]. . . . So we stayed in close touch with Marjorie . . . I admired what she was doing. (Barry 2012)

Barry had also worked on a number of civic projects with Harry Lester and Andrew Fine, two of the founders of LRN, and spoke at length with them about their efforts to address water quality issues in the Lynnhaven River watershed. Having lived on the Nansemond for some twenty years, Barry was interested in preserving the natural beauty of the river (Barry 2012). Barry began to talk to neighbors and friends about the idea, and found that there was a great deal of interest in such a group. Smith, one of the early members of NRPA, relates that she ran into Barry at a local gas station:

> Dick Barry was there putting gas in his car and I was there and he said, "Karla, let me talk to you a minute." He said, "I'm interested in starting an environmental group for the river. Would you help me?" And I said, "Well sure, just let me know." And so, a few months later, he called me and said that we should get some people together and then we met (Smith 2012c). Dick Barry asked people he knew, that he knew were concerned about the environment; who were doers; who would really do something. (Smith 2012b)

The initial group consisted of about fourteen people, nearly all of whom were invited by either Barry or Smith. Two of the more influential early members were Jack Eure and Howard Martin, both of whom grew up on the water in Hampton Roads and who had first met Barry in law school at the University of Virginia (Eure 2012). The early meetings focused on the development of a mission for the nascent organization. There were a number of different interests represented by the participants. Several wanted to try to limit future development in the city generally, and along the river specifically. Others were concerned with water-quality issues, and still others were interested in education. Smith raised the issue of access to the waterway, and suggested to the group that the environmental message would resonate more with citizens if there was greater public access—"Why would somebody downtown want to contribute to an organization that's cleaning up the river they can't get in?" (Smith 2012c).

The initial slate of officers was developed by a combination of consensus and good-natured arm-twisting. There was general consensus that Barry should be president, but many of the others had less of an interest in becoming officers. Smith wanted to be on the education committee:

> So then Dick [Barry] said, "You need to be the vice president." No, I want to be on the education committee. He said, "No, you get to do both . . ." And then we couldn't get anyone to be treasurer, so then . . . I think it was Dick, said, "John [Wass], you need to be treasurer." [Wass] said, "I'll be treasurer until we get the first dollar and then I'm not doing it because I don't do money." And he hasn't gotten off it yet. . . . And then we talked Alice Mountjoy into being secretary. (Smith 2012b)

The next issue for the group was to develop a mission statement. The nascent education committee, Smith and Wass, had developed a mission statement for that committee as a means to express to the rest of the board the purpose of the education committee. In an extended conversation that required three meetings to complete, the board ended up "cannibalizing" (Smith 2012b) the education committee's mission statement. The end result was a statement that captured the different interests within the group, and that the group "could all live with and felt good about" (Smith 2012b). The final version of the mission is as follows:

> *The mission of the Nansemond River Preservation Alliance is to raise public awareness and encourage environmental stewardship of the Nansemond River, Chuckatuck Creek, Bennett's Creek and their tributaries and wetlands by developing and implementing programs and initiating collaborative actions with individuals, businesses, civic organizations and governmental agencies.* (Nansemond River Preservation Alliance 2012b).

The next step in the formation of NRPA was to look for more permanent leadership. Initially, Barry had asked a long-time friend, Wilson Browning, to serve as a part-time executive director. Browning served in this capacity for about a year, but had other interests. On August 30, 2010, NRPA received its 501(c)(3) nonprofit status from the Internal Revenue Service (IRS). The application for the tax status had been prepared by a law firm located in Charlottesville, Virginia, but Martin, an attorney in Norfolk and resident of Suffolk, handled the creation of the organization's bylaws, incorporation documents, and other associated paperwork. Barry and Martin had been in law school at the University of Virginia together, and Barry asked Martin to assist with the paperwork: "He's our attorney. Doesn't get paid anything. You know we said, 'You're going to do some lawyering and you're going to do it for free'" (Barry 2012). Martin saw an opportunity to help both an old friend and his community:

> I went to law school with Dick Barry. Dick has been a friend of mine for many years, and he's the one who called me up and asked me if I would be willing to be on the board and to assist them with some of their legal items that might come up, including the organizing entity. I have done some pro bono work—that means work without necessarily getting paid for it . . . I do have experience in doing a lot of work with governmental entities. (Martin 2012)

Prior to IRS approval of its nonprofit status, NRPA needed to devise a method to allow individuals to donate to the organization, but still enjoy the benefits of the tax laws governing charitable giving. Barry approached Christy Everett, the director of the Chesapeake Bay Foundation's (CBF) Norfolk office, and the state director, Ann Jennings, to discuss the issue:

> You want people to be able to take a tax deduction when they give you money because we're now a 501(c)(3) tax exempt organization . . . on Day One you don't have that. So we asked them if they would be our fiscal agent so contributions could be made to the Chesapeake Bay Foundation and they would funnel it through to us. And, of course, they were very careful about that to make sure that the money they gave us was being used for the purposes. That was a big help. You know, people would say, "Send the check to the Nansemond River Preservation Alliance, care of Chesapeake Bay Foundation." That was a big help. (Barry 2012)

This seemingly minor arrangement provided an important incentive for people to support the new organization financially, and thus allowed NRPA to begin its work almost immediately. The arrangement ended when NRPA was granted its 501(c)(3) status in August 2010.

Around this time, Elizabeth Taraski was approached by Barry about joining the organization. Taraski had a significant career in fundraising, and had most recently been at Christopher Newport University in Newport News, Virginia, as vice president for development. She had recently married and moved to Suffolk, and was interested in getting more involved in the community (Taraski, Smith, and Wass 2012). When she was referred to the group, Barry approached her about joining NRPA, and he did so in a typical low-key fashion:

> She wasn't looking for a big, heavy duty job. She liked to play golf sometimes. Her husband [had] retired. . . . We said, "Look, would you like to help us out? Maybe a fundraiser, next year become an executive? All part time. Money's terrible." And she said, "Okay, I'll do it." (Barry 2012)

Taraski began in a limited role in October 2010, and became executive director in March 2011, still on a part-time basis (Taraski, Smith, and Wass 2012). Among her first tasks were efforts to begin a marketing campaign, raise funds for the organization, and grow the membership. She headed up the effort to create a website for the organization, and to build alliances in the community. She began a series of meetings with, and presentations to, other civic groups in the community, as well as representatives of the City of Suffolk. Often she was accompanied by committee chairpersons from NRPA, and the chairperson would provide a brief presentation on the current efforts of NRPA. Nick Worth, the chairman of the water quality committee, accompanied Taraski on a series of meetings at which he presented data on the state of water quality in the river:

> She and I had a dog and pony show. . . . We'd go around and talk to clubs and you-name-it. I have a little slide show, and people would get riled up when they found out about the [water-quality issues in the] upper Nansemond. One gentleman jumped up and said, "You mean I

> can't catch fish?!?" And I said, "I'd catch and release, buddy—can't swim in it; I wouldn't eat out of it." (Worth 2012a)

The NRPA held its first River and Creek Fest on June 2, 2012, at Bennett's Creek Park. The event featured displays contributed by both NRPA and several of their partner organizations. One display featured Chesapeake blue crabs; visitors could watch as crabs shed their shells and began to form new shells. Another display contained an aquarium full of murky water; several oysters were added to the tank, and the water would clear in about an hour.[6] Other displays contained information on home oyster gardening, rain gardens, and native plants. The local Girl Scout group provided a display on the importance of recycling.

The other popular activity of the day was seeding oysters on NRPA's demonstration oyster bed. Several members brought boats, and Volvo-Penta provided both boats and life jackets. Small oysters had been collected from schools and citizens participating in the citizen oyster-gardening program—about thirty thousand oysters in all. Children were taken out to the reef on the boats, where they seeded the bed with the oysters. Interest was high; the boats shuttled kids back and forth to the reef for about four hours. An estimated 150 people attended the event during the day.

COLLABORATION PROCESS

The mission statement of NRPA highlights the importance of collaboration for the group: "and initiating collaborative actions with individuals, businesses, civic organizations and governmental agencies" (Nansemond River Preservation Alliance 2012b). The importance of collaboration was echoed by nearly every person interviewed for this study, many of whom referenced the mission statement directly. There was also widespread agreement that a collaborative approach is much more desirable than a confrontational approach:

> I think the goal is not to be confrontational. That's how we started out. I don't think anyone in the organization wants to be confrontational. We want to work it out first . . . try to work it out . . . realize that both sides have their wants and desires. And that's the idea. (Wass 2012a)

> I think it's a really good approach . . . because, one thing it does is it forces you to understand the perspective of the other side so you can't just see them as the enemy or the evil developer or whatever. You have to get to know them. You have to get to know their objectives and goals. You have to get to know their kind of currency, whatever it is they deal in and speak to them in the same language. You learn a lot that way. As with most things it's rarely what you thought. People are coming from a different place than you often anticipated. Oftentimes people are perfectly willing to collaborate and perfectly willing to

> change certain things about their project. They're not cutthroat or ruthless. Whenever there is that opportunity, if you have a collaborative group, you have your foot in the door versus if you're taking the hard line it's going to be a lot more difficult. (Gonzalez 2012)

Stakeholders

Like the other two groups in this study, the NRPA began with the actions of a small group of people. The first task for NRPA was to identify a core group of stakeholders who could be enticed to join the fledgling organization and contribute to its growth. As detailed earlier in this chapter, Barry reached out to people he knew in the Suffolk community to build the initial core membership. Unlike our other two case studies, however, the initial efforts to build a core group focused almost exclusively on personal relationships, with little thought to the inclusion of important stakeholder groups. This does not represent a fault as much as it does a difference in approach. While the founders of our other case study groups immediately focused on questions of stakeholders, Barry and others approached the formation of NRPA as though it were a civic group. The people who were invited to the initial meetings were people in the community who had a reputation for getting things done in the community—people with connections to other people and other civic groups, and who could be trusted to put their energies into the nascent organization.

Indeed, to the extent that other stakeholders have been invited into the group, many have voluntarily sought out NRPA, rather than being sought out by the organization. Bob Crocker, the operations manager at the Volvo-Penta Test Center located on Chuckatuck Creek, heard about the group from a friend in the Rotary Club; this person was also a neighbor of Barry (Crocker 2012). Geoff Hinshelwood, a member of the Wetlands Board, learned of the group through members' attendance at Wetlands Board meetings, and decided to bring his daughter to a meeting one evening (Hinshelwood 2012). Cindy Linkenhoker, a stormwater engineer with Parson-Brinkerhoff and resident of Bennett's Creek in Suffolk, learned about NRPA when she visited the Ruritan Hall for a local art exhibition and saw a sign advertising an upcoming meeting. Linkenhoker had served as the stormwater manager for the City of Portsmouth for nearly fifteen years, during which time she worked closely with ERP on a number of projects. She understood the value of these organizations, and was pleased to learn about a group in her community (Linkenhoker 2012). Steve Barnum, a recently retired executive with the National Oceanographic and Atmospheric Administration (NOAA) and a Suffolk resident, heard a talk given by Taraski at a meeting of the Suffolk Master Gardeners:

> We had a representative come from the Nansemond River Preservation Alliance. And she spoke to us and I went up to her and talked to her after the presentation. I told her I'd like to learn more; I'd like to join because I live in Suffolk, you know, the Nansemond River, although I don't live on the Nansemond River . . . people ask if I live on the water and I say no; I bought the house on the highest ground I could find . . . and so I know what storm surge does. (Barnum 2012)

Another factor that dictated this approach was the relative lack of clearly identifiable stakeholders in the watershed. With little industry on the water, there were few identifiable potential business partners. Likewise, the traditionalistic nature of the city, and the "arms-length" posture of the city government, meant that the city was largely disinterested in the group at its outset. Two other potential stakeholder groups, farmers and watermen, have viewed NRPA with something akin to cautious suspicion. Both groups have historically been fiercely independent, and have been wary of any efforts related to environmentalism. While some inroads have since been made with the Nansemond watermen, those efforts have been due to personal relationships rather than organizational appeal (Smith 2012c). In terms of water quality, the agricultural community is an important stakeholder group:

> I think they're going to have an uphill battle. Some of that comes from the fact that it's an agricultural community—and it's a personal frustration I'm sure they'll face too—the agricultural industry is not regulated for stormwater runoff . . . in the entire state. And it is the single largest polluter sending about 40 percent of the sediment load to the Bay every year . . . that's been a thorn in my side for a long time. . . . You're talking about people's food now . . . the way the state does it is they give cost share money through soil and water conservation districts—then the districts work with farmers to encourage them to do voluntary controls—if they put in the controls, they can get the money. Some are going to do that, but a lot aren't. So I think there's going to be difficulty for NRPA to get past that hurdle. . . . If you really want to get to clean water, then at the end of the day—somehow agriculture needs to be brought into the discussion. (Linkenhoker 2012)

Another important stakeholder group that was not represented at the founding of the group was the real estate developers. Although many of the interviewees for this study identified overdevelopment as the most important threat to the watershed, there is little evidence that the group sought participation from the developers. There may be several factors that account for this condition. First, many of the developers who have undertaken projects in Suffolk are not located in the city; rather they are located elsewhere, buy land to develop, and then begin the zoning and permitting process in Suffolk to move forward with their plans. Thus, it is difficult to know a priori whom to invite to the table. Second, environmental groups are often considered the bane of developers, and developers often go to great lengths to avoid local environmental groups. Third, NRPA was formed during the national housing slump in the latter part of the decade, and new construction in Suffolk had slowed considerably because of a lack of demand (Taraski 2012c). Many of our interviewees noted that the role of developers would become much more important as the housing market strengthened. Finally, some in NRPA felt that they were not yet sophisticated enough in their knowledge to be effective advocates for the river, and that they didn't want to engage the developers until the group had a better understanding of the state of the watershed and its issues. When asked about this issue, Barry (2012) said, "We're not ready yet. We have to get our learning curve advanced. We need to know more about this than anybody else."

The NRPA is still a young organization, and continues to seek out new members and partners. In the past two years, NRPA has made an effort to make contact with other organizations in the community, and has worked to build relationships with the local chapters of the Izzak Walton League, the Nansemond River Power Squadron, the Suffolk Master Gardeners, and others (Taraski 2012a; Worth 2012a). Taraski (2012b) also noted that local civic leagues have not yet been brought into NRPA, and this will be a focus of future efforts to expand NRPA's partnerships. Additional efforts have been made to develop additional relationships with the City of Suffolk, particularly through city departments. To date, NRPA has established good working relationships with the departments of Public Works, Planning, and Economic Development. At least two members of the city council have attended NRPA events (Barry 2012), and a candidate for mayor in the 2012 election attended the inaugural River and Creek Fest in June 2012, as well as the presentation of the River Trails map held in October 2012. As the group gains greater acceptance and recognition in the community, it is reasonable to expect that additional relationships and alliances will be formed with additional stakeholder groups in the community.

Conveners

There are several individuals who serve as conveners for NRPA. The initial convener was Barry. An unpretentious man with a self-effacing sense of humor, Barry was the motivating force behind the initial formation of the group, and he used his position in the community and his contacts to gather a group of people to form the organization. Drawing on his experience in other community projects and efforts (Barry 2012), Barry identified the core individuals he thought necessary to build the momentum of the group, and persuaded, cajoled, and convinced these individuals to join. Many joined simply because they were asked:

> So I talked to friends, neighbors out there . . . just kind of hinted around to see if there was any natural interest. And there was. It amazed me that there was spontaneous interest in it. . . . There's a ready market for this kind of activity . . . pent-up demand. (Barry 2012)

Barry also provided financial resources in the early years of the organization to provide operating expenses until the group began to seek membership and grant resources. However, Barry wants the group to evolve past a reliance on him as a leader:

> I don't want [NRPA] to be dependent on me. It was in the early days, and it's becoming less so. I want that evolution to continue. And we want strong people coming along who will be eligible to take my place. . . . That's my corporate experience. You can't build any organization on one person. So that's very much a goal of mine so I can just say, "It's your turn now," and I'll take a back seat in the room. (Barry 2012)

Another important convener for NRPA is Smith. Smith is, in many ways, the central node of the network that ties NRPA to the local community, particularly the long-time residents of Suffolk. While Barry describes himself as "Norfolk-centric" (Barry 2012), Smith has lived in the community for decades, and is a self-described "doer" (Smith 2012b). Her involvement with the local historical group, SRH, adds to her position as a convener capable of turning out members and volunteers for NRPA activities. Moreover, her long history of civic activism, coupled with a friendly, outgoing personality, allows her to motivate friends and acquaintances (as well as strangers) to become involved in civic projects.

A third convener of some note is Taraski, the executive director of NRPA. As the public face of NRPA, Taraski spends a great deal of time talking to other local groups, as well as the media. In a relatively short period of time at NRPA, Taraski has made numerous important contacts in the community, and has worked to build the membership of NRPA. Just as important, however, is her internal focus on the effectiveness of the organization. Taraski attends all of the committee meetings, and serves as a conduit of information between the committees and the executive board. She works with the committee chairs to plan events, meetings, and activities, and helps ensure that the committees have the resources to conduct their programs. Although these activities are what one might expect of an executive director, they are also consistent with the definition of "convener" discussed in chapter 2.

Resources

The NRPA is still a relatively new, and consequently small, organization. Financial resources in the early days of the organization were provided largely by donations from the initial group of founders, and the organization's budget has grown as both membership and grant activity have increased. For 2010, the first year in which NRPA filed a Form 990 with the IRS, the group reported income of $25,635, against total expenses of $12,209. The 2011 Form 990 reports $80,216 in revenue, and total expenses of $74,994. At the end of 2011 NRPA reported net assets of $18,648.[7] The numbers for 2012 will be higher, reflecting both increased membership and additional grant funding. Aside from the cash account, the only tangible resources of the group include computers, office supplies, and some water-quality testing equipment. Some of the testing supplies were purchased with grant funds, but the equipment—dissolved oxygen meters, refractometers, and Secchi disks—were purchased by Nick Worth and Dick Barry and donated to NRPA.

There are other categories of resources for which NRPA fares better. Although NRPA has only one paid employee (the part-time executive director), the 2011 Form 990 lists seventeen members of the governing body and 110 volunteers. Indeed, the participants interviewed for this study underscored the ability of NRPA to mobilize volunteers to participate in events such as the oyster demonstration reef, the citizens' oyster-growing program, water-quality testing activities, and the social and education events hosted by the group. Members of the study team attended three such events—two River Talk events at the Ruritan Hall and the River and Creek Fest held at nearby Bennett's Creek Park. At the River Talk events, volunteers arrived early to set up chairs in the meeting space, set up the projection and sound systems, and prepare the kitchen space. Other volunteers arrived, many of whom brought food placed on a large table at the side of the room. Others served as impromptu parking attendants, directing those arriving for the program to designated parking areas. At the end of the programs, many attendees stayed behind to talk to friends, and nearly everyone helped clear the floor of chairs and tables, clean the floor space and kitchen, and ensure that the room was clean and tidy. During the River Fest, volunteers were present to talk about NRPA with visitors, answer questions about the river, help organize people involved in the oyster-planting activities, load boats with baskets of spat (baby oysters), and clean up the area at the end of the day.

When asked whether the group's political resources were sufficient, our interviewees expressed different opinions. Most agreed that NRPA was seen by the community as legitimate, and capable of addressing water-quality issues in the river. The degree of coverage in the local newspaper would tend to support this observation; the media coverage to date has been exceptionally positive. When asked specifically about the city government, however, respondents provided a range of answers. An anonymous respondent expressed optimism that the city government would take NRPA more seriously, and several hoped for a close working relationship with the city:

> All the bureaucrats in the city are just scared to death. They don't want anybody getting them in trouble. They don't want to hear about some newspaper story that came out where somebody said something bad about them. We're trying to say, "We're not going to do that to you. We're here to help you!" So we're getting through that part.

Other respondents pointed to ongoing collaboration with the city, including the Water Trail mapping project, participation on the Wetlands Board, and a series of small projects with the Department of Public Works. Indeed, there seems to be something of a bifurcation of support in the city. At the agency level, NRPA has made some solid connections with individual city administrators, and has developed close working relationships with several key people. In addition, at least two members of the city council have taken an active approach to NRPA, and have been publicly supportive of their efforts. The NRPA has been less effective to date in engaging the more senior city administrative officials. The management style imposed by the current city manager is best described as "risk averse" and "insular." Policies imposed by the city manager make it difficult for citizens and others to approach city offices; in fact, requests for direct contact with city officials must first be approved by the Office of Media and Community Relations. As part of this research, we contacted this office and requested permission to talk with three city employees; our request was never approved.

Rules

As a small, community-based organization, NRPA tends to be very informal in its approach to its activities. As a 501(c)(3) entity, the organization has formal bylaws they follow. An analysis of the bylaws reveals that the main thrust of the bylaws is to create a governance structure of specific offices and duties for its official positions. The bylaws are largely unremarkable, and appear to contain fairly standard language for the purpose. Conspicuous in its absence, however, is any mention of collaboration. Section 1.1 of Article 1, "Specific Purposes," presents the organization's mission:

> The primary purposes of the Corporation are to preserve and improve the environmental quality of tidal waterways, tributaries, rivers, creeks, and watersheds in or touching Suffolk City and Isle of Wight County in the State of Virginia, specifically including the local portions of the James River, all of the Nansemond River, all of Chuckatuck Creek, and all of Bennett's Creek. The Corporation may embark upon all legal activities furthering the environmental quality of the rivers, creeks and tributaries of the tidal and watershed areas located in Suffolk and/or affecting the waterways mentioned above. The Corporation may undertake any and all reasonably necessary or desirable activities

> assisting in the support of these purposes. (Nansemond River Preservation Alliance 2010)

To the extent the use of collaboration becomes a rule within NRPA, it is thus an informal rule.

Along with the definition of formal positions within the organization, the bylaws also specify four initial committees: the governmental relations committee, the education committee, the governance committee, and the executive committee. As detailed elsewhere in this chapter, the education committee was a specific focus of at least two of the initial members of the board, Karla Smith and John Wass. The bylaws allow for the formation of additional committees, and two other standing committees (the water quality monitoring committee and the shoreline/wetlands committee) have been added to date.[8]

Meetings held by NRPA are as much social events as they are working meetings. At a meeting of the shoreline committee held in April 2012, for example, many of the members arrived early, and spent their time greeting their fellow committee members. The group was welcomed briefly by Taraski, the executive director, who then turned the meeting over to John "Jack" Eure, the committee chairperson. Eure is a tall, gentlemanly, soft-spoken man who thanked all of the attendees, and then introduced the two presenters for the evening. There was a perfunctory approval of the minutes from the previous meeting, and the presentations began. The committee members asked questions after each presentation, and a free-flowing discussion developed. Eure did not need to formally recognize speakers; the group was respectful and all were allowed to speak without interruption. Several of the attendees took careful notes during the meeting. Though there was a formal agenda for the meeting (and that agenda was followed), there was no mention of an agenda during the meeting. The meeting was notable for its lack of formality. The meeting adjourned an hour after it began, although many of the attendees remained to socialize with others.

The process by which officers were chosen, described earlier in this chapter, is also indicative of the informal nature of the group. The combination of volunteerism, persuasion, and arm-twisting was good-natured, and the participants were all willing. Eure, chair of the shoreline committee, described the process in this fashion:

> . . . [laughing] . . . I think I got "Shanghaied." I don't know . . . maybe Dick [Barry] or Elizabeth Taraski . . . they just asked me to do it. I don't think I was elected by any kind of majority voting. I just ended up with it. I don't really have any expertise on the shoreline preservation. . . . I probably know less than my committee people but I just happen to be the committee chair this year. . . . But I'm learning. (Eure 2012)

The organization is new enough that it is still developing informal rules. However, there are two notable informal rules that have permeated the organization so far. The first concerns contact with the media. The board of directors decided that, with the appointment of an executive director, all requests for comment by the media would be referred to the executive director, and that no member should speak on behalf of the group without prior consent (Smith 2012b). The second rule is known within the organization as the "Dick Barry Rule." In short, the rule is that no meeting may last more than one hour, regardless of whether the business of the group is finished or not (Smith 2012b). When asked about the origins of this rule, Barry explains that he developed the idea during his career as a business executive. He feels that meetings without time limits encourage people to engage in needless discussion, and to impose a time limit keeps the group on point. If the issues are too complex to be decided in a single meeting, then they are carried over to the next meeting. Keeping the meetings short is also a means to keep people interested, and helps people feel as though their commitment to the organization is not needlessly open-ended. The "Dick Barry Rule" is almost religiously observed at all organizational meetings, and many of our interviewees made note of the rule during our discussions.[9]

Governance Structures

The NRPA is a small organization, with only one part-time paid employee. The structure of the organization is thus completely reliant on a combination of a committee arrangement (driven largely by board members), coupled with volunteers drawn from the organization's membership. The development of the committee structure in NRPA was an important element of the early formation of the organization (see figure 6.2). The original group discussed the use of committees, and debated ideas for the different committees that might be formed. Smith and Wass, two of the early participants, were retired teachers, and had a strong interest in the inclusion of an education component in the organization. The two formed the oyster restoration committee, and focused almost immediately on oyster gardening and oyster demonstration projects as the focus of the education efforts. Smith had been involved in similar projects during a thirty-nine-year career as a teacher, and understood the lasting impact of on-the-water experiences for children. She had been bringing children to her house on field trips since 1998 to raise baby oysters. Both Smith and Wass knew how to navigate the school district bureaucracy, and who to approach for funding for education programs. The education mission began almost immediately, and a demonstration reef was established just northeast of the Godwin Bridge, on about an acre of river bottom donated by a local resident (Taraski, Smith, and Wass 2012). Its location near the mouth of the river assured adequate salinity, dissolved oxygen, and low bacteria counts in the water, as well as plenty of water flow due to tidal action in the river. The oyster restoration committee now boasts three NRPA members, along with fourteen school group leaders representing public and private schools in Suffolk, and two representatives of the home schooling community. It has also spun off the Citizen Oyster Restoration group, with another fourteen members (Nansemond River Preservation Alliance 2012c).

The other committee that began in the early days of the organization is the water quality committee. The first head of the organization, Wilson Browning, had a technical background, and focused his attention on issues of water quality. Nick Worth, a retired executive with Landmark Communications and one of the early members of NRPA, assumed responsibility for the water quality committee. One of the first tasks was to determine the state of the river. Worth sought historical data from the Virginia Department of Environmental Quality (DEQ), as well as the Virginia Department of Health, the Hampton Roads Sanitation District, the Hampton Roads Planning District Commission, and the CBF. Between November 2010 and spring 2011, Worth and his committee collected, collated, and analyzed the available data on water quality in the river (Worth 2012a). Worth credits Chris Moore, a scientist with the CBF's Norfolk office, with guidance on how to both analyze and interpret the large amounts of data the water quality committee had at their disposal, including the use of a geometric mean to smooth the data and make long-

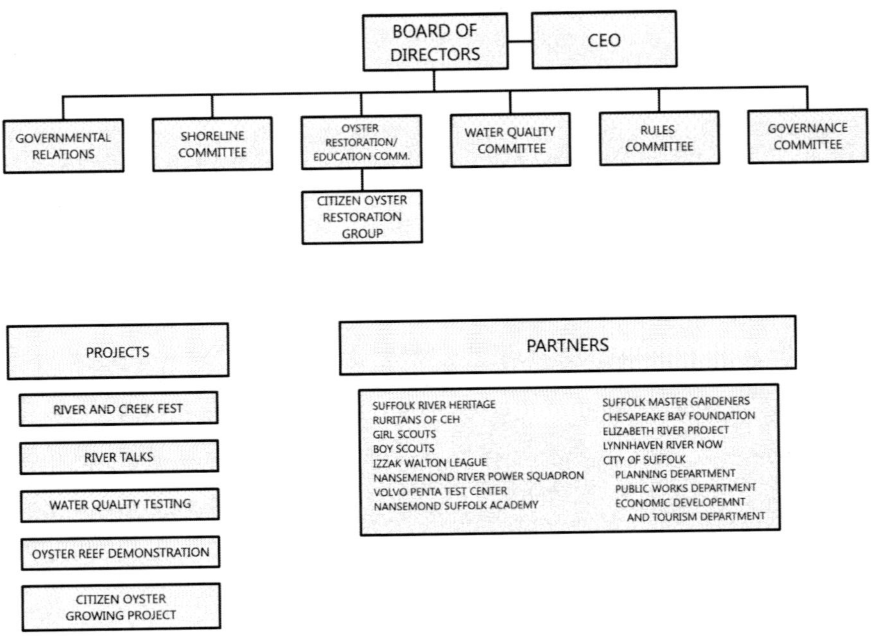

Figure 6.2. Organization Chart—Nansemond River Preservation Alliance

term trends more detectable (Worth 2012b). In June 2011, the committee prepared and released its first State of the Nansemond report. The committee had adopted a "report card" format similar to that used by both LRN and the CBF, a format that reduced the data to a letter-grading system to make the message more clear to citizens. The report showed that while the lower Nansemond was in reasonably good shape, the upper Nansemond (in downtown Suffolk) was heavily impaired, with high levels of phosphorus and nitrogen, and high bacteria levels (Nansemond River Preservation Alliance 2011). The report was released in June 2011 at a presentation held in downtown Suffolk at a hotel built on the banks of the river, and was covered extensively in the region's largest newspaper (Harper 2011e; *Virginian-Pilot* 2011).

The next step for the committee was to assume responsibility for water-quality testing on the Nansemond. Due to funding constraints, the DEQ had stopped water-quality testing on the river, but had created a citizen monitoring program. Worth and his fellow committee members applied for participation, completed the required training, and received certification to begin monitoring in late 2011; they began collecting samples in January 2012 (Worth 2012a). The committee collects data once a month, collecting samples and taking measurements at seven sites, from the mouth of the river to the upper reaches of the river near Constant's Wharf in downtown Suffolk. They test for dissolved oxygen, bacteria levels, water clarity, and salinity. Measurements for all but bacteria levels are taken onsite, and samples are tested for E. coli bacteria after the collection trip. The data are loaded into DEQ's database by committee members (Worth 2012a). Much of the equipment for the water-quality testing efforts was bought through small donations from members, although both Worth and Dick Barry contributed significantly by purchasing two dissolved oxygen meters (Worth 2012a)—an expensive device that must be regularly calibrated. Bob Crocker at Volvo Penta, an important ally of NRPA, often donates a boat and fuel to the group to travel to the testing sites. On some days, Worth offers the use of his boat and fuel (Crocker 2012). The group received a grant of $4,000 in 2011 to offset some of the costs for expendables, fuel, and so on, and they hope to be able to renew the grant in future years (Worth 2012a).

Members of the research team for this book accompanied the water-quality committee members on their testing trip in June 2012. The boat was a thirty-six-foot aluminum test craft used by Volvo Penta to test marine engines, transmissions, and electronics systems, and was captained by Crocker. In addition to Worth, two other water-quality committee members were on the team. After leaving the Volvo-Penta facility on Chuckatuck Creek, we headed east and then southwest into the Nansemond proper. When we reached the first test site, the boat was stopped, and the committee members began their process. First, small plastic bottles were attached with rubber bands to the shaft of an old golf club; this was lowered into the water to a specific depth to capture a water sample. Small drops of water were placed on a refractometer to measure salinity levels. Several readings were made with both dissolved oxygen meters, and the temperature of the water was recorded. The oxygen readings were corrected for water temperature, and readings from both instruments were recorded on data sheets along with salinity levels. Finally, a weighted white-and-black disk, called a Secchi Disk, was lowered by a thin line into the water; the testers measured the depth at which the disk was no longer visible to the unaided eye. This test measured water clarity. Each stop for collection and measurement took about fifteen minutes to complete. The entire trip lasted about three and a half hours.

The most recent committee to form is the shoreline committee. Initially, the prevailing thought was that shoreline preservation should be part of the water quality committee, since water quality is closely linked to the existence of buffers and riparian zones along the waterway. Nick Worth, chairman of the water quality committee, realized that the shoreline issue was too complex and important to be relegated to an existing committee, so he recommended to Dick Barry that the organization form a separate shoreline committee (Worth 2012a). The executive committee agreed, and the committee was formed in January 2012. Headed by Eure, a Suffolk attorney, the committee is charged with ensuring compliance with the Chesapeake Bay Preservation Act (CBPA) that, among other things, seeks to preserve wetlands and vegetative buffers. The Suffolk planning commission is charged with implementation of the act, which is accomplished through the actions of the Wetlands Board of the commission. NRPA has sought to build and maintain a close relationship with the planning commission and its Wetlands Board, and several NRPA members serve on the commission or the board. The Wetlands Board in Suffolk dates back to 1972, and the city adopted ordinances to implement the CBPA in September 1990 (these were slightly amended in 2003).

One such member of the Wetlands Board and NRPA is Geoffrey Hinshelwood. Geoff is the co-owner of a water-quality testing laboratory in Hampton, Virginia, and has lived in Suffolk for about eight years. He was appointed to the Suffolk Wetlands Board in 2009, and joined NRPA in 2012. Hinshelwood attended NRPA's River Talk in May, at which Walter Priest, a well-known authority on wetlands restoration from the Virginia Institute of Marine Science, gave a talk about the process of wetlands restoration:

> I thought, well I'll go see Mr. Priest's demonstration. I was very impressed with not only him, but the folks I met there . . . and I brought my twelve-year-old daughter and wanted to get her involved more than I wanted to get involved. It turns out that both of us decided we'd better volunteer and try to get involved—we live right around the corner from Bennett's Creek, so it seemed like a natural fit. (Hinshelwood 2012)

At the shoreline committee meeting on April 26, 2012, the guest speaker was Brian Alperin from the City of Suffolk Planning Department. There were twelve NRPA members in attendance, and the meeting began with introductions of all those present—the committee was new enough that not everyone on the committee knew each other. The group listened attentively as Alperin described the city's efforts to implement the CBPA. There were several thoughtful and insightful questions about both technical matters and matters of policy. One of the more critical questions addressed the required setbacks from the water's edge to any structures on the property. The CBPA specifies a minimum setback, and this is codified in city ordinances. However, several members of the committee questioned whether the setbacks were adequate to protect the river, and whether the Planning Commission was allowing developers to build inside the setbacks. Gretchen Gonzalez (2012), a biologist who recently joined NRPA and serves on the shoreline committee, explained the issue:

> [to achieve the full effect of setback] theoretically, you'd have to buy up all the land around the water to a certain distance out—just buy it up and protect it. Reforest it, replant it . . . that's not realistic and I'm not saying that's what somebody should do . . . [the appropriate buffer zone] depends on your site, you know. It might be two thousand feet; it might be you don't even need one depending on what the site is.

Rather than take an adversarial position with the Planning Commission, NRPA seeks to adopt a model similar to what they believe to be the case with the ERP. They foresee a scenario in which they would work with both the Planning Commission and developers, so that developers would come to NRPA before seeking approval from the Planning Commission. NRPA would work with developers to ensure that the developer's plans are environmentally friendly, and consistent with both city ordinances and the CBPA. Rather than oppose the commission or the developers, NRPA sees working collaboratively with these groups as its role, with the goal to allow development to coexist with protection of the river. Eure describes the goal as follows:

> We took the position early on that we needed to work with [the city]. I think that collaboration is a very good word for that. Instead of being a watchdog or a critic we try to work with them to accomplish mutual goals. . . . We've tried to assure others in the city that, "Hey, we're not out to create trouble for you. We're out to try to work with you to accomplish cleaning up the river . . ." We've heard that on the Elizabeth River Project that they've gotten to the point where when . . . an industrial developer, wants to do something along the Elizabeth River, the planning department refers those developers to the Elizabeth River Project and says, "You go see those folks and see what they think and they can help you and give you some guidelines. If you make them happy you'll probably make us happy . . ." What a remarkable thing, if you've got that kind of a status. We're not seeking that kind of political

> power but if the city and the Wetlands Board had enough faith in what we're doing . . . it may be that we become their allies. We would love to be there . . . [we're] not there yet. (Eure 2012)

The two other committees in NRPA, the governance committee and the governmental relations committee, appear to be more ad hoc in their activities. While the governance committee was reasonably active in the very early formation of NRPA, its current role is to be supportive of the operational committees (education, shoreline, and water quality). As chair of the governance committee, Martin ensures that the activities of the committees are in accordance with the organization's bylaws, and he provides advice and guidance as required (Smith 2012b). The governmental relations committee, chaired by Eure, maintains connections with the City of Suffolk. Eure's office is located across the street from City Hall in downtown Suffolk, and his role is generally to be a point of connection between NRPA and city officials (Taraski, Smith, and Wass 2012; Martin 2012).

Collaboration and NRPA: A Summary

The NRPA is a small organization with limited resources, but with a great deal of spirit and common purpose. Although the group is still finding its role in the community, the notion of collaboration is firmly imbedded in the emerging culture of the group. As a new, citizen-based collaboration, they have little in the way of tangible resources to offer in a collaborative setting; when they work with other groups in the community, NRPA offers a vision, legitimacy, and legions of volunteers. To date, most of the collaborative efforts have involved NRPA making efforts to engage with other groups in the community. An important exception to this is the partnership with SRH. In this case, SRH, through the efforts of Smith, reached out to NRPA and offered assistance and connections. The shared office space in the Ruritan Hall, initially the domain of SRH, is illustrative. Likewise, the Suffolk Water Trail map project was very much a collaborative effort. The map project clearly had benefits to both groups, and by combining their resources they were able to leverage additional resources from the City of Suffolk and the National Park Service to produce a product of significant use to both groups.

More recently, NRPA's work with schools in Suffolk has produced additional collaboration successes. Teachers from the Nansemond Suffolk Academy have been involved with NRPA's oyster-education program since its outset, and have formed a close partnership with NRPA as a result. Recently, the Nansemond Suffolk Academy applied for, and received, funding from the Virginia Environmental Endowment to expand their oyster-education program. As Jerry McCarthy, the executive director of the endowment, explains,

> [NRPA has] a relationship with a group called Nansemond Suffolk Academy—a middle and high school I guess it is . . . that we also gave a mini grant to recently, and that's only because they're working with NRPA as one of the schools that NRPA had identified as a partner—and now they've built other schools as well. . . . For a small group starting from scratch, they're moving at warp speed it seems to me—and still managing to do it effectively. And what I think we started helping them with at first was the oyster-gardening idea—and we made a second grant, I think aimed more at the educational stuff. (McCarthy 2012)

The grant to the Nansemond Suffolk Academy will allow NRPA to shift its resources to other schools, while still working with the Nansemond Suffolk Academy students and faculty to support and tend to NRPA's oyster demonstration reef.

COLLABORATION OUTPUTS AND WATERSHED OUTCOMES

As a new organization, NRPA has little in the way of identifiable outputs, and even less in terms of watershed outcomes. However, there are several outputs that can be identified that have resulted from the group's activities and collaboration with others. Unlike the ERP, where the nature of the problems in the watershed require long-term plans and projects, NRPA has not yet developed a Watershed Action Plan. Just as the problems on the Nansemond are less complex, the projects undertaken by NRPA are less complex.

Outputs

There are three broad categories of outputs present in NRPA. The first of these is water-quality monitoring. The water quality committee was formed shortly after the formation of NRPA, and an early task was to determine the current state of the river. With the cessation of state monitoring, there was no clear picture of the state of the watershed. As noted earlier in this chapter, the first task of the committee was to obtain previous water-quality data from the Virginia DEQ, and to use those data to produce a "report card" on the state of the river. This was accomplished in mid-2011, and the report was released to significant media attention. The water quality committee continues its monitoring activities, and plans to produce an updated report every two years. These data will not only inform citizens about the state of the river, but will also provide the City of Suffolk with information relevant to its responsibilities under the TMDL (Total Maximum Daily Load) requirements now in place for the Chesapeake Bay and its tributaries.

The second category of output is education. As with our other two case studies, education programs are a critical part of the mission of

NRPA. The importance of the education component was noted by Smith, a retired teacher, and co-chair of NRPA's education committee:

> You can't have an organization built with [age] sixty and above membership. You've got to make it go to young adults, and my other thing is you've got to capture the enthusiasm of the high school, middle school, and younger people so that they are in that engagement of environmental stewardship. So that it's natural that they'd be a part of this group when they got older. So that's an opportunity in Suffolk that no one's ever had. (Smith 2012b)

An example of the effects of environmental education was offered by Janice Gates. A long-time resident of Virginia Beach and a member of both the ERP and LRN, Gates recently moved to Suffolk and joined NRPA:

> I couldn't get my own child in his thirties to recycle, but when his daughter came to see me—we would recycle—then she put her foot down and said, "Dad, we can't just throw this stuff away." So, if there's something going on in the schools, kids can be agents for change. (Gates 2012)

The main thrust of NRPA's efforts to date has involved its oyster-restoration program. There are currently eleven schools in Suffolk involved in the program, and the number continues to expand. The group succeeded in creating an oyster demonstration reef, and is considering the establishment of demonstration reefs in other parts of the watershed (Wass 2012b). An effort in 2012 to involve local Suffolk students in the group's first River and Creek Fest fell apart due to miscommunication within the school district (and concerns over liability and safety), but the school district has since solved those issues and has pledged its participation for the next event (Smith 2012b).

The NRPA's involvement in education goes beyond the oyster-restoration program. Its "River Talks" are open to the public, and are chosen to help educate citizens about the river. At one such talk held in May 2012, Walter Priest, a well-known expert on wetlands restoration, gave a talk on the issues of restoration design and construction to a group of about eighty-five attendees. Other talks have addressed the history of the Nansemond watershed, the work of groups like LRN and the ERP, and other similar topics. Nearly all of the displays at the River and Creek Fest were educational in nature, and ranged from how to plant riparian buffers with native vegetation to oyster gardening to the effects of pollution in the river. As noted elsewhere in this chapter, the presentations made by NRPA members to other organizations in the community are largely educational in nature, as the group seeks to generate interest in efforts to preserve the watershed.

The third category of output to date involves efforts to expand access to the river. The nature of the Nansemond is such that nearly all of the available shoreline is privately owned, either by farmers or by individual homeowners. As noted earlier in this chapter, the issue of access was important in the initial discussions of the mission of NRPA. Smith (2012b) explained the importance of the issue of access:

> Now we did put the John Smith Water Trail out to everybody because CEH [now Suffolk River Heritage], our group on the Chuckatuck, had started that and I wanted the Nansemond River [Preservation Alliance] to collaborate because I thought it would make a stronger case. So I had presented a whole thing about the John Smith [Water] Trail. What were the pros and cons, where we wanted to go with it and the mapping thing and the fact that it would promote public access. . . . At the beginning, Dick Barry couldn't understand why public access should be in our mission statement. And I said, "Well, Dick, if you have 98 percent of the people who own a whole shoreline of Suffolk and only 2 to 5 percent even have anywhere where they can get access to the river, and there's only one boat ramp, how are we going to convince the other 95 percent that they should be in the Nansemond River Preservation Alliance when they can't use it?" So if you don't have public access, then this organization is only meeting the needs of the shoreline landowners, who are wealthy . . . so you're not addressing the inner city of Suffolk; you're not addressing the Bennett's Creek neighborhoods; you're not addressing inland areas. And I said, "It's all going into the watershed. You have to expand what you're doing to include everybody and we've got to have public access. In events, at those projects, it then includes the people of Suffolk." And then he understood why public access should be part of what we were doing. . . . Otherwise it's a yuppie club, like a yacht club. You won't last long.

The Suffolk Water Trails map was released in October 2012 at a River Talks event. The map project is a prime example of collaboration between organizations. Growing out of a realization that access to the river is an enduring issue for the Nansemond, the map is designed to raise awareness and interest in the river, and to help grow ecotourism in Suffolk. While the project began as a collaboration between SRH and NRPA, NRPA was able to secure an in-kind grant from the National Park Service (NPS) to integrate the project into the NPS's Captain John Smith Chesapeake National Historic Trail. A multiyear project, the NPS seeks to map and commemorate Smith's voyages to explore the Chesapeake Bay between 1607 and 1609. Smith's travels took him to the headwaters of the Nansemond River, and the NPS was pleased to have a group take an active interest in the project. The NPS provided an in-kind grant of staff time to provide technical assistance to the project. The story was picked up in the local paper, which led to an inquiry from the city's Department of Economic Development and Tourism. The city quickly realized the benefits of the project, and agreed to provide resources to print the map. While there were some early disagreements about how much control the city would exercise over the project (Smith 2012b), the collaboration was otherwise smooth.

The map, measuring eighteen by thirty-two inches, is printed in full color. One side of the map is dedicated to historical locations on the river, and marks some eleven historically significant sites on the river. Other sections trace the explorations of Captain John Smith on the lower Chesapeake Bay, discuss the history of the Nansemond tribe, the role of African American oystermen on the Nansemond, and a short box on the famous Civil War battle between the *Monitor* and the *Merrimac* in 1862 (which took place just a few miles from the mouth of the Nansemond River). The other side of the map lists the few places on the Nansemond, Chuckatuck Creek, Bennett's Creek, and reservoirs and lakes, created when the river was dammed, that offer access to the waterways. In addition, there are boxes with information on wetlands habitats, and the flora and fauna found in the watershed. At the bottom of the map is a list of the map's sponsors (nine in all), as well as QR (Quick Response) codes for both Suffolk River Heritage and NRPA, which allow those with a smartphone to scan the code and connect with the respective organization's websites. The map will be distributed at several locations in the city, by the participating sponsors, and will be available at kiosks in select locations in the watershed.

Outcomes

Because of the relatively young age of NRPA, there are no measureable outcomes attributable to the group. While the oyster-restoration program has been implemented and the demonstration reef is being monitored, the reef has been in place less than a year. Likewise, it is too early to know whether the education programs in place will serve to create changes in environmental ethics and behaviors among the participants. Anecdotal evidence from participants suggests that behaviors will change, but these will likely manifest themselves over a period of years. As NRPA becomes more established in the community, we might expect to see longer-term effects related to buffer zones in new housing developments, as well as greater efforts by citizens to restore natural shorelines. Changes in water quality are likely to be minimal for the foreseeable future,[10] due to the lack of water flow in the upper reaches of the river.

SOCIAL CAPITAL

As noted in chapter 2, social capital is both an output of the collaboration process, as well as a necessary precondition for collaboration to occur. The NRPA provides evidence of the role of social capital in both of these places in the model; this discussion will illustrate the degree to which social capital becomes the "glue" that binds citizen-based collaboration.

Social Capital as a Precondition for Collaboration

As noted in chapter 2, a certain amount of social capital must exist in a community in order to make citizen-based collaboration possible at the outset. Without an initial degree of trust, legitimacy, and commitment to place, there is little incentive for citizens to engage in collaborative endeavors, particularly in an arena (or a cause) to which they may have only an indirect connection.

In the case of NRPA, there was already a significant amount of social capital present in the community, particularly in the northern part of the city. More importantly, Barry recognized the importance of tapping into existing social capital. Although he described northern Suffolk as a "hodgepodge" and a bedroom community for Norfolk and Portsmouth (Barry 2012), he also clearly understood the importance of social capital:

> What happens is people get involved and they just love it. They just love it because the community loves it and everyone gets associated with it, which is exciting. It's a little bit infectious from that standpoint. There's your social capital—there's a ready market for this kind of activity—pent-up demand. (Barry 2012)

The initial ties with the local heritage association were also instrumental in providing initial social capital for NRPA. As delineated earlier in this chapter, Smith was effectively the bridge between the heritage group and the new environmental organization, and thus also a bridge between the "come heres" (Smith 2012b) and the long-time residents of the area. Because Smith was trusted by both groups, she was effectively able to provide a ready source of legitimacy to the fledgling group, and the trust she engendered in both communities made the integration of the participants a very straightforward process.

Elements of Social Capital

The elements of social capital—trust, norms of reciprocity, legitimacy, and efficacy—are closely tied together. The responses from our interviewees illustrate the interconnections:

> —I think legitimacy is important, because that's the foundation for trust. (Worth 2012b)
>
> —The city recognizes [NRPA] as a trustworthy group . . . we can get things done [efficacy]. (Hinshelwood 2012)
>
> —We're filling a need out there for individuals that very much have an interest in the water and seeing Suffolk move forward and progress. (Taraski 2012a)

—You know, you do stuff for other people because you're asking them to do for you [reciprocity] . . . you do this with people you trust. (Smith 2012a)

By reaching out to other citizen groups in the community, NRPA has been able to tap into existing reservoirs of social capital in the community. As noted earlier in this chapter, NRPA members regularly meet with other groups and give presentations on the state of the river and the work of NRPA. Participation by other groups in NRPA events has also strengthened social capital in the community. At the River and Creek Fest in June 2012, many of the displays were provided and staffed by other civic groups, including the Suffolk Master Gardeners and the Girl Scouts.[11] The oyster seeding relied significantly on the participation of local Girl Scout troops. An effort in 2011 to place medallions on storm drains in Suffolk is another example. The city provided the medallions, and NRPA facilitated the project, and local Boy Scout troops installed the medallions on stormwater sewer covers (Taraski, Smith, and Wass 2012). This level of interaction not only helps to provide a sense of legitimacy for the group, but it also serves to engender trust and efficacy.

Efficacy is a particularly important element in the furtherance of social capital, in that it provides evidence to the community that the group can accomplish its goals. While NRPA is still a very young organization, the efforts of NRPA's three "action" committees—the water quality committee, the education committee, and the shoreline committee—are noticed by the community. The NRPA has received very positive coverage in the local press, particularly for its water-quality report card. The very act of water-quality monitoring is an overt example of action by the organization, and the favorable reaction of the community to the report builds a sense that NRPA can make a difference. Time will tell whether NRPA is able to translate this efficaciousness into a sustained ability to both grow the organization and to change behaviors in the watershed, but the early efforts have shown very positive returns.

When asked whether NRPA's activities have helped to increase the amount of social capital in the community, our interviewees were uniformly sure that NRPA has had a positive effect on the amount of social capital present. Many pointed to the rapid growth in the organization's membership; others pointed to the group's philosophy of engagement (as opposed to confrontation):

> We're not going to be that group that walks with signs in front of a company. Our whole thing from the very beginning, from the very first meeting, was "We want to be a group with solutions; we want to see what we can do to maintain what we have already in the City of Suffolk and our waterways." And if there are problems we don't want to be the one pointing them out—the whistleblower—we want to be the one that looks at the problem, and . . . see what we can do to help

> them . . . [determine] the solutions that can happen on a local level without being on the front page of the newspaper. I mean, nobody wants that. (Smith 2012b)

A wonderful example of reciprocity was provided by Crocker, the manager of the Volvo Test Center on Chuckatuck Creek. Smith has been actively engaged with the Girl Scouts in Suffolk for many years. She and her husband, Jimmy, are avid sailors, and are the owners of a scale replica of a traditional Chesapeake skipjack, a wooden fishing boat powered primarily by sail.[12] When Hurricane Isabel struck the area in 2003, the boat was sunk at the dock behind the Smith's house on Chuckatuck Creek:

> She's an amazing woman . . . I'm in awe. In fact, her sailboat sunk in [Hurricane] Isabel . . . and she'd been teaching these Girl Scouts how to sail. So I got her a new Volvo engine [donated by Volvo Penta] for her sailboat . . . and we take care of her because she's just extraordinary. People like that are great. (Crocker 2012)

Impressed with her commitment to her community, Crocker offered his resources to help Smith continue her community work. When NRPA was formed some years later, Crocker saw an opportunity to reciprocate, and he has given generously of his time and resources to support NRPA. This, in turn, has served to encourage other NRPA members to do the same, which in turn builds additional social capital.

Commitment to Place

As detailed in chapter 2 of this book, and in the preceding two case studies, commitment to place is not simply an element of social capital; it also serves as a potent motivating force that spurs individuals to participate actively in citizen collaborations. Nearly every person interviewed for this case study expressed a long-standing and powerful commitment to the water. For some, the connection began on the Nansemond:

> I was born and raised actually two blocks from here and then when I was about seven years old my parents moved to a section called Riverview and it was, in fact, at the time on the river. That part of the river has since been dammed to form Lake Mead but when I was a kid we were called "river rats." We played and the river was our backyard. We were muskrat hunting and duck hunting and fishing and sort of "Tom Sawyers" up and down what was really a creek then, the Nansemond River. And we just loved it. And then later when I was fifteen years old my grandfather gave me what's called a Barbour, wooden lapstrake boat, with a thirty-five horse motor and that was the biggest thing going at the time. And a bunch of the kids around town had that brand of boat and so then we started water skiing and venturing up further and down the river. Then a group of us used to go, three or four summers in a row, we'd actually take the boats and team up with two

> people per boat and we'd run all the way to Virginia Beach and through the Lesner Bridge, Lynnhaven river, Crystal Lake. That was a big deal when you were sixteen years old. That was quite an adventure. I grew up enjoying the water. My home now, we live on Chuckatuck Creek. That's more of my immediate involvement because I still love the water and want to preserve it in a lot of ways. (Eure 2012)

Others came to the area later in life, but saw in NRPA a means to take an active role in their community:

> I live in Suffolk: this is where I live; this is where I bicycle; this is where I buy my groceries and everything. . . . And I thought, well, the Nansemond River is in my backyard and I want to help. So I just volunteered. (Barnum 2012)

More importantly, there is an inherent recognition among members of NRPA that others in the group share the commitment to place. As Taraski (2012c) said,

> Social capital is very important . . . we didn't form in response to any type of urgent detrimental event that was happening in Suffolk. It was different groups of individuals coming together with a common interest. . . . It's the water. So many of these people have boats or they fish or they just live on the water . . . so it's the water. I feel that it's people's place to preserve and restore [the river] for future generations also. You had to have the opportunity to enjoy it. I've heard so many stories about downtown Suffolk from people who used to go swimming in that water . . . everyone is delighted that someone is taking a leadership position in protecting the waterways.

One group that seems to remain largely outside the growing sphere of social capital is the local watermen. The NRPA is working actively to bring the watermen into the group, but it is something of an uphill battle:

> Yes, I think, and we don't have them yet; we're working on them. We've got to get the watermen behind this 100 percent and they're beginning to be nice. But they're sort of, like, skeptical about what we're doing. And I think . . . the word "tree hugger" has been thrown at me a few times, and that's fine. Most of the people in Nansemond River Preservation Alliance are not your local watermen and local villagers, and so they're looking at this group of "come heres," that's what they like to call it, who are a little more affluent . . . and I have to say that a lot of times, the group that is advocating for environmental action is a little more educated, a little more affluent, and for . . . they don't want this group of people to come into their waterways where they've been throwing a beer bottle overboard for twenty-five, thirty, one hundred years, and tell them what they can and cannot do. And so there's a little pushback from that. I'm in a rare place, because I'm friends with all the watermen. And I want them to help me with the oyster thing and they're arguing with me about the concept of oyster restoration for restoration's sake and using tax money to do it. And so I had to argue,

> no, our oyster restoration is for educational purposes, and we're not using tax money to do it, we're using grant money that is not always federal. (Smith 2012b)

This observation from Smith illustrates a fundamental tension between those who make their living from the water, and those that seek to preserve the waterway. Chesapeake watermen have a well-deserved reputation for independence and a historical distrust of government,[13] and they tend to view environmental groups as a threat to their livelihood. In addition to being an important stakeholder group for NRPA, they have also resisted the efforts to bridge across communities and thus further build social capital. Smith realizes the long-term importance of extending the bonds of social capital to the watermen, and is confident that NRPA will be successful in bringing the watermen into the group:

> I'm slowly winning them over. And Jimmy, my husband, is helping me a little bit. . . . And he's caught between his friends and me wanting to help. I think I'm going to win them over. I think they're starting to be nicer to me. But it's political, I mean, it's socially political in this community. I don't know when we're going to win over the rank and file neighborhood people. One family at a time. But once they're all on board, I think [NRPA] be a force that will be hard to deal with. You know, they're not putting us down, they're just not running to join. I don't know how to describe it; you have to tread carefully. (Smith 2012b)

SOME FINAL THOUGHTS

The story of NRPA provides an example of the ability of a grassroots environmental organization to identify its strengths, and to build on those strengths to propel the nascent organization forward. There are several lessons that emerge from this case; the highlights are listed in textbox 6.1.

Textbox 6.1

Lessons from Nansemond River Preservation Alliance

- Bringing stakeholders into a collaboration is a process, not an event.
- Education programs are a means to reach both children and their parents.
- Having a clear mission and goal is critical for long-term success.
- Trust between participants is enhanced by participation in the collaborative process.

- Boundary spanners can help connect groups in a community.
- Education is a vehicle to bring future members into the organization.
- Initial conveners are critical to group success.
- Recruiting members of other groups in a community can help bridge groups and strengthen social capital.
- The more vibrant and active organizations in a region, the more social capital will develop between collaboratives. Collaborations will share lessons learned with other collaborations.
- Work with other local citizen groups can leverage existing social capital in a community.
- The connection to place is a powerful motivator for participation.
- Citizen-based collaborations do not just create social capital; they use social capital to accomplish their goals.
- Successful collaboration requires a central convener figure.
- The nature of the problem in the watershed will determine who the stakeholders are and the goals of the collaboration.
- The availability of intellectual capacity and expertise is likely to lead to a greater success for the collaboration. Grassroots organizations must attract and retain experts to volunteer their time.
- One of the most important factors leading to successful collaboration is the ability of participants to "check their egos at the door."
- Projects allow retired people to contribute to worthy goals. This is the first generation of retirees with environmental knowledge who are available to volunteer their expertise in retirement.

The NRPA has accomplished a great deal in a short span of time. As a relatively new organization, they have not been in existence long enough to see the outcomes in their watershed that have been seen in the Elizabeth and Lynnhaven Rivers. However, NRPA has benefitted greatly from the experiences of both the ERP and LRN, and they have been able to structure their group around the successful lessons from their sister organizations. Like the others, NRPA places a great deal of emphasis on collaboration and consensus building, and they largely reject a philosophy of confrontation.

The long-term challenge for NRPA is to define a specific set of goals to achieve. Unlike our other two cases, NRPA's mission statement does not seek to accomplish a goal by changing the current state of affairs; they seek to maintain the status quo. The lack of definable goals will likely make it difficult to sustain interest among both citizens and collaborative partners, as there will be no reasonable means to measure the success of the group's efforts. Whereas LRN can point to the percentage of the river open to oyster production, and ERP can advertise its successful restoration projects, NRPA's goals are much more modest. An important difference between our case studies is that both ERP and LRN developed measureable goals very early in their formation process, while NRPA appears more focused on activities.

Likewise, NRPA's efforts to influence development along the Nansemond River are yet to bear fruit. The most important in this arena will likely be the group's success at building rapport with city government, particularly the Planning Department and its Planning Commission. Without a firm and supportive relationship with the city, NRPA will have very limited influence with developers. The NRPA will likely find it necessary to develop independent connections with developers, if they seek to bring developers into a collaborative structure. If NRPA is seen by developers as an adjunct of the Planning Commission, there will be greater opposition by developers who may see NRPA as an impediment to their business opportunities. If NRPA is committed to collaboration (and all indications suggest that it is), then it must work to find common ground with developers and demonstrate to developers that collaboration with NRPA is worthwhile. Similarly, efforts will be needed to draw both farmers and watermen into collaboration with NRPA.

On the other hand, NRPA is an excellent example of the power of social capital in the formation of a citizen-based collaboration. The physical nature of the community, along with the prevailing political culture, provides a veritable incubator for the growth of social capital. While not every community is endowed with these attributes, the case provides some important clues for relevant factors in similar communities. The ability to tap into existing social networks provides a ready source of social capital, in that trust, norms of reciprocity, and commitment to place are likely to be present. Moreover, a commitment to the new organization by respected people in the community can lend legitimacy to the effort, which can also build confidence in the efficacy of the group.

As this chapter was being drafted, a pressurized sewer line that crosses a small tributary ruptured, spilling an estimated one thousand gallons per minute into Shingle Creek (Sheler 2012). The leak continued for more than two weeks before work crews were able to reroute the sewage flow and repair the twenty-inch concrete sewer line. It is not known at this time what effect the spill will have on the short-term health of the river, although increased bacteria counts are likely to close sections of the river to shellfish harvesting. Incidents such as this suggest that threats to the Nansemond River are not limited to suburban encroachment and nonpoint source pollution, and also serves to underscore the importance of the water quality monitoring efforts of NRPA.

NOTES

1. Committees are especially important in NRPA and LRN, because these two organizations have relatively small staffs. On the other hand, ERP has several full-time employees, so much of the implementation work is carried out by staff, rather than committee members.

2. Among other things, the context in which NRPA operates is very different. The organization is also much newer and less well developed; many of its programs are just being developed. These issues are more fully discussed later in this chapter.

3. Bulkheading is the process of building a wall of wood or steel at the point at which the water meets land. Because bulkheads are typically built so that the bulkhead extends below the low tide level of the water, the natural wetlands of the shore are lost as the vegetation at the shoreline is now completely separated from the water. Any water that runs off the land thus runs directly over the top edge of the bulkhead and into the waterway.

4. Estimates by the Virginia Employment Commission suggest that the population of Suffolk will reach nearly 151,500 by 2030 (City of Suffolk 2012).

5. Previously known as the Crittendon, Eclipse, Hobson Historical Foundation, the organization formally changed its name in October 2012.

6. Oysters eat by filtering bits of algae and other organic matter they ingest and then expel. This action also takes minute bits of sediment and other debris out of the water. A medium-size oyster can filter about five liters of water per hour (Judy 2012).

7. Data from the Form 990s in this section were obtained from Guidestar, an online repository for information on nonprofit groups. Information reported on Form 990 is, by law, public information. The data may be located at www.guidestar.org.

8. While the bylaws list one of the standing committees as the "Education Committee," and NRPA members refer to the committee by this name, the NRPA website lists the committee as the "Oyster Restoration Project Committee." Likewise, the NRPA members refer to the newly formed committee to address shoreline issues as the "Shoreline Committee," but it is listed as the "Wetlands Committee" on the website.

9. Several interviewees also asked, tongue-in-cheek, whether the "Dick Barry Rule" applied to the interviews we conducted during this research!

10. The rural nature of northern Suffolk means that there are a great number of houses with septic tanks. Beginning several years ago, the city now requires all residents with septic systems to have their septic tanks pumped out every five years. While this policy predates the NRPA, the group has been actively working with the city to test for (and address) issues of broken or leaking sewer lines (Taraski, Smith, and Wass 2012).

11. The Lafayette River Fest in April 2012, sponsored by the ERP, followed a very similar pattern. While the ERP had two tents set up with displays, there were nearly fifty other tents and displays provided by other citizen groups, local government agencies, and local environmental companies.

12. Skipjacks were traditionally powered only by sail. In the early twentieth century some skipjack captains began to carry a small boat equipped with a small engine to serve as an alternate source of power when needed. Smith's skipjack, a scale replica, was built with a small inboard engine in place of a powered dinghy. Traditional skipjacks are rigged with a very large mainsail and a smaller jib (foresail), and have a very shallow draft to allow the boat to fish in shoal areas. While depths in the Bay can reach over one hundred feet, the average depth of the Bay is about fifteen feet (Warner 1976).

13. For an excellent discussion of the historical independence of the Chesapeake watermen, see Warner (1976). Likewise, Blackistone (2001) touches on these themes in a series of interviews and verbal portraits of watermen from the region.

SEVEN

Conclusion

One Watershed at a Time

The easiest way to eat an elephant is one bite at a time.

—Robert Crocker (2012)

On the surface, grassroots collaborative efforts might appear to be simple, but they are hardly that. Finding workable approaches to "wicked" environmental problems requires sustained effort and focus, which is a difficult undertaking for any citizen-based, grassroots collaborative. This book has examined three cases of grassroots collaboratives involved in the restoration of three watersheds in southeastern Virginia, as well as an overview of region-wide efforts for the Chesapeake Bay. It should be noted that nonprofit organizations are not in and of themselves collaborations, but use collaboration as a means to accomplish goals. The genesis of these organizations was groups of friends and acquaintances that shared a common interest. These nonprofits serve as focal points for actions to clean up the rivers and provide connections to allow people to take action. None of the organizations in our case studies have played the "blame game." They all have recognized that trying to find who is at fault for the condition of the rivers is not useful, especially when the damage to the rivers occurred decades previously.

This chapter serves to summarize the major findings of our research. Because we seek to help inform other grassroots efforts, we offer the findings and lessons in textbox 7.1 in the hope that these bits of knowledge will be of use to other groups who wish to make a difference in their communities. We begin with a brief review of our three case studies, and then discuss the overall findings of our work. We conclude with a reexamination of the five themes we articulated in the introduction to this volume.

Textbox 7.1

Overall Lessons

- The more citizens organize themselves, the greater the level of social capital and stewardship generated in the community.
- The greater the diversity of stakeholders involved in a local collaboration, the greater the breadth of voices that can speak to policy makers, and the greater their credibility with local, state, and federal officials.
- The presence of other grassroots organizations in an area provides an opportunity to build social capital and encourage sharing between organizations.
- Having group members who are also members of other civic groups can help connect groups in a community.
- Committee structure matters. Committees should clearly reflect and support the group's mission.
- Successful collaboration requires respect for diverse opinions and a willingness of participants to treat others with respect.
- Education is a critical component of the work of grassroots organizations. Education is a way to bring future members into the organization, and can help change behaviors of both children and their parents.
- Context matters. The nature of the problem in the watershed will determine who the stakeholders are and the goals of the collaboration.
- Conveners are a critical resource in the formation and operation of the collaborative effort.
- Social capital matters. Grassroots collaboration relies on initial social capital in a community to begin a partnership, and successful collaboration can help build additional social capital.
- Grassroots collaborations do not "own" environmental problems; the responsibility rests with the community. Grassroots collaborations provide a vehicle through which communities can work together to address environmental problems.
- Expertise matters. Grassroots environmental organizations can draw on retired professionals to provide needed expertise.
- A commitment to place is a powerful motivator. A positive emphasis on place—Because It's My Backyard—can inspire people who otherwise might not get involved.
- High-profile champions can enhance the chances that the initial formation of a grassroots collaboration will be successful.

- It is important to invite all stakeholders to the table. Not all may participate, but all efforts should be made to be as inclusive as possible.
- Partnerships with local government can greatly enhance the resources available to grassroots environmental groups, and can provide increased legitimacy for both the citizen group and for government.
- Grassroots environmental groups can make a measureable difference in the watershed.

A REVIEW OF THE CASES

Our three cases exhibit many similarities, yet they are also different in many ways. Using the model introduced in chapter 2 as a guide, we now turn to a brief comparison of the three cases.

Context

All three cases are located in the greater Hampton Roads area of Virginia, and are thus identical in terms of the broader national, state, and regional contexts. Since all three watersheds drain into the Chesapeake Bay, all three rivers have been subject to previous (and current) efforts to address water quality in the Chesapeake estuary, inasmuch as any small tributary has been included in those efforts. All three watersheds are subject to the Bay Total Maximum Daily Load (TMDL) requirements, and the political jurisdictions in all three watersheds face many of the same challenges in meeting the TMDL requirements.

The local context of each of these cases, however, differs in some important ways. First, in spite of a common history and a shared broad political culture, the political culture in each watershed exhibits subtle differences that, in turn, help determine the ways in which local grassroots environmental groups are received by both citizens and governments. Three of the cities in the Elizabeth River watershed, Norfolk, Chesapeake, and Portsmouth, tend toward a more individualistic culture. Although the watershed extends into the western portions of Virginia Beach, the City of Virginia Beach has largely been ignored by the Elizabeth River Project (ERP), primarily because ERPs focus has been on the southern branch of the Elizabeth, and little attention has been paid to the eastern branch of the river.

In contrast, Virginia Beach tends more toward a blend of moralistic and individualistic cultures, primarily due to city administrative leadership. Virginia Beach city government has been much more willing to engage in open, participatory approaches to governance, and citizen engagement has been a cornerstone of this approach. The net result is a city that is much more focused on quality of life issues, with a particular flair for quality of life improvements that include all citizens. This is evidenced by the zeal with which the city initially engaged Lynnhaven River NOW (LRN), a relationship that has continued to build and strengthen over the years. LRN's success is due, in no small measure, to the willingness of the city to support the work of LRN, financially, politically, and in the allocation of resources.

These two watersheds stand in stark contrast to the community surrounding the Nansemond River. While the cities of Chesapeake and Virginia Beach have longstanding agricultural roots (as does Suffolk), the explosive population growth seen in these cities has not yet occurred in Suffolk. Farming interests are still strongly entrenched in Suffolk, and this difference is visible in the strongly traditionalistic political culture of Suffolk. While more recent trends suggest that Suffolk is the next high-growth area in the region, the political culture in the city will likely change more slowly than its population numbers.

The nature of the problem in the three watersheds also differs. The Elizabeth River is a traditional industrial river, and has been subjected to centuries of environmental abuse. The problems in the Elizabeth watershed are severe, complicated, and expensive to remediate. ERP's more recent interest in the Lafayette River, a tributary to the Elizabeth, broadens the scope of ERP, as the water-quality problems in the Lafayette are largely due to residential runoff. Still, ERP's focus remains fixed on the industrial pollution found in the southern branch of the river. This focus necessitates the inclusion of industrial interests along the river, a group not traditionally known for its affinity for environmental groups.

The Lynnhaven River suffers from explosive residential growth, and the concomitant combination of stormwater runoff, legacy septic systems still found in parts of the city, and runoff from lawn fertilizer, pet wastes, and sewage discharge from boaters. LRN has settled on a broad spectrum of approaches and programs designed to address these nonpoint issues, many of these aimed at social marketing, alteration of citizen behavior, and efforts to upgrade both stormwater and sanitary sewer infrastructures. The Lynnhaven River is the most densely populated watershed in our study, and there are no shipping or heavy industrial operations in the watershed.

The Nansemond River differs from the other two watersheds in some significant ways. First, it is the least populated of the three watersheds. There are many active farms situated along the river, interspersed with housing developments. Man-made dams at the upper end of the river have significantly altered the natural hydrology of the river, which in turn has exacerbated nonpoint runoff problems, particularly at the more densely populated upper reaches of the river. Unlike the other two watersheds, public access to the Nansemond is almost nonexistent, which means that most residents of Suffolk have little chance even to see the river, much less enjoy the river. The Nansemond River Preservation Alliance (NRPA) has made river access a cornerstone of its early efforts. Like the Lynnhaven River, there is little in the way of industry located on the river.

Each of our three groups began with little in the way of financial resources. The initial efforts of each group were largely supported by donations from the founders. However, the nature of the problems in the watershed meant that ERP began to seek grants very early on, a strategy that continues to this day. LRN pursued a mixed strategy, combining citizen membership with significant financial support from the City of Virginia Beach and, more recently, grants from organizations such as the USACE, VIMS, and others. While it has sought some relatively small grants, the NRPA currently relies very heavily on citizen memberships and donations to support its activities.

Social capital also played a very significant role in all three cases. For both LRN and NRPA, social capital in the community was a key component to their early success. Both organizations relied on a network of friends and acquaintances to provide early momentum, and both benefitted greatly from the early involvement of well-known and well-respected citizens in their respective communities. ERP faced a different challenge, in that the founding members were passionate, but lacked the influential contacts in the community. Moreover, ERP realized that their social capital was not citizen-based, but rather organizationally grounded. The founders of ERP thus had to work to build legitimacy for their efforts, for which they relied heavily on their ability to build personal relationships with other stakeholders in a mutually supportive, nonthreatening manner.

Collaboration Process

All three of our case-study organizations have successfully employed collaboration to achieve changes in their respective watersheds. Each group has identified the important stakeholders in the watersheds, and has worked actively to incorporate these stakeholders into the collaborative structure. Their governance structures are very similar, and they have all relied heavily on committee structures to set goals, identify projects, and implement plans. The central organizations are all 501(c)(3) nonprofit entities, and thus have bylaws and organizational officers. ERP and LRN have specifically sought stakeholders to serve on their committees, a strategy that assures that stakeholder input is made a part of the governance and decision-making processes. The NRPA is still new enough that it has not yet made significant efforts to reflect all stakeholders in the governance structures, a task that our other case studies suggests is a critical component for long-term success.

All three groups also espouse a collaborative, nonconfrontational approach to their work. This is especially important for two reasons. First, this approach serves to build trust between participants. The businesses located along the Elizabeth River know that ERP's motive is not to place blame, or embarrass its partners into action. Rather, ERP has positioned itself as a trustworthy partner. Second, businesses have found that association with ERP has a spillover effect that lends legitimacy to its efforts at environmental compliance (as well as its voluntary efforts). Businesses now take pride in their voluntary efforts, and are working more diligently to maintain full compliance with regulatory requirements.

All three cases underscore the importance of conveners in collaborative settings. Whether they are founders, executive directors, or members, each of these cases has at least one person who serves as a facilitator (McNamara and Morris 2012) to activate the group, frame the process and operating rules, mobilize participants, and synthesize an environment conducive to positive interactions between participants (Agranoff and McGuire 2001). Conveners can also be organizational; our examination of the Bay-wide setting suggests that groups like the Chesapeake Bay Foundation (CBF) or the Choose Clean Water Coalition (CCWC) can serve to being people to the table, and act as conduits to share information and expertise between groups.

Outputs and Outcomes

All three of our case-study groups have identifiable outputs. In our study, the age of the group is directly tied to the breadth and depth of outputs, as well as the existence of measureable outcomes. The NRPA, the youngest group in our study, has initiated a water-quality testing program, an oyster-gardening program, and an educational program. Its relatively young age, as well as the modest scope of its early goals, precludes the existence of outcomes. On the other hand, both ERP and LRN have a multitude of outputs. These include "green" recognition programs for homeowners, schools, and businesses. ERP's River Stars program is particularly notable, in that the program has become a source of pride to its participants. Both the River Star Homes (ERP) and Pearl Homes (LRN) programs have a large membership, and new participants are constantly joining. All three case-study organizations have invested much in their educational efforts, again with great success. ERP's Learning Barge is a highly visible indication of its commitment to education, and the oyster programs sponsored by LRN and NRPA draw hundreds of students each year.

Both ERP and LRN can lay claim to measurable changes in environmental quality in their watersheds. While it is clear that the groups cannot claim sole credit for the changes, it is equally clear that much of the improvements would not have happened "but for" the efforts of these groups. Each group employed collaborative processes to marshal expertise, money, support, and social capital to make the projects happen. LRN's success is indicated by a tremendous increase in the amount of the river now open to oyster production—from less than 1 percent of the river at the time of the group's founding, to 40 percent of the river as of this writing. The gains for ERP have been somewhat less dramatic at a watershed level, but are nonetheless remarkable given the nature of the problems in the watershed. The cleanup efforts at Money Point have resulted in an extraordinary revitalization of the ecosystem at Money Point, one that has far surpassed the expectations of experts. Similar gains are likely to be seen at Paradise Creek Park as future studies assess the changes in that portion of the river. These successes seem to defy conventional wisdom that grassroots environmental collaborations are ineffectual at meaningful environmental change.

Social Capital

All three groups have not only relied on social capital to fuel their initial formation; their work has helped to build additional social capital in their respective communities. For LRN and NRPA, the gains in social capital have occurred mostly at the citizen level. These groups have been supported by individual citizen members, and their success at enlisting new members and volunteers for projects is testament to their success at building social capital. Both groups are seen as good stewards of the environment, and they have reputations as reasonable, thoughtful, and mainstream organizations. They have built both trust and legitimacy within their communities, and are seen as capable of addressing water quality concerns in their communities.

ERP has also built social capital, but in a different venue. Because so many of its stakeholders are business and industrial concerns, ERP has worked hard since its inception to build trust and legitimacy among this stakeholder group. This has been a particularly challenging task, since there are stakeholders with a longstanding animosity toward environmental groups. ERP's model of a nonconfrontational, "no-fault" approach, coupled with an ingenious incentive/recognition program, has paid handsome dividends. Many of its industrial partners have not only undertaken voluntary projects, but some have actually mentored other (sometimes rival) businesses to create new voluntary projects. This is not only a setting in which businesses do not want to be left out of the recognition that participation brings, but it also spawned a friendly competition in which businesses attempt to "one-up" their neighbors with new projects. This has not only built significant trust for ERP, but has also engendered trust and collaboration between business rivals.

In sum, our three case organizations are much alike in some respects, and very different in other respects. The differences in processes, outputs, and outcomes can be largely attributed to differences in context, particularly in the challenges facing each watershed and the political culture in which the watershed is located. The similarities in structure, operation, and social capital are attributable in part to a commitment to collaboration (a structure in which everyone participates voluntarily, in which everyone participates as an equal partner, and in which decisions are made collectively), in part to the legal requirements connected to 501(c)(3) status, and in part to a deep commitment to the community. They are all working for a shared vision of a better community, and taking a pragmatic approach—as we said early in this book, they are not "Dark Green" or "Light Green"; they are interested in what works. Most importantly, they have succeeded where previous efforts have failed.

THE THEMES REVISITED

In the introduction to this book, we identified several themes to explore in this research: the centrality of social capital; a commitment to place; the importance of context; the potential benefits of collaboration between grassroots citizen groups and stakeholders; and the ability of grassroots collaboratives to actually make a difference. Although the preceding discussion touches on these themes, a brief review of the themes helps underscore the importance of these elements, and helps frame both guidance to other grassroots organizations and larger lessons for scholars interested in grassroots collaboration.

Social Capital

Our three case studies, and our discussion of the interactions between our case-study organizations, highlight the significance of social capital in grassroots collaborative efforts. There must be at least a minimal amount of social capital available in a setting in order for people to agree to collaborate. Grassroots collaboration can lead to greater levels of social capital and stewardship in the community. Social capital is grounded in trust and reciprocity between citizens, and successful collaboration requires trust and reciprocity between participants. The trust-building process must occur in a deliberate, nonthreatening environment for successful partnerships and social capital. Whereas agency-based collaborations begin with relatively low levels of trust, citizens seem inherently to trust their fellow citizens. Essentially, nonprofit organizations are not seen as having "hidden agendas," a factor that contributes to trust. Clearly, partnerships require trust to be effective. The more transparent a nonprofit organization is in terms of its mission, processes, and decision making, the greater the level of social capital that can be generated. Trust between participants is enhanced by their participation in the collaborative process. The existing level of social capital at the outset provides an important basis for action, and can serve to bring others into the collaboration. Having more stakeholders involved in a local collaboration increases the breadth of voices that can speak to policy makers, thus increasing their credibility with local, state, and federal officials. Social capital translates to political capital.

Closely connected to the concept of social capital is the idea of stewardship. The three organizations studied in this book do not "own" the problem or the solution to the problem of rivers at risk. It is incumbent upon them to work with all stakeholders since stewardship is everyone's responsibility. The projects undertaken by these organizations have provided significant opportunities for retired people in the community to contribute to worthy goals. This is the first generation of retirees with environmental knowledge who are available to volunteer their expertise in retirement. Education is an important aspect of stewardship. In fact, education has been a part of almost all the projects of these organizations even if not given a high priority for any particular project. Programs in schools get stewardship commitment from the next generation. In many cases, it is the children who are teaching their parents the behaviors to model. When children care, parents care, and when homeowners care, neighbors care. Education programs can help convince children that the current state of the rivers is not the normal state of the rivers. Education can help to build stewardship and social capital across economic class lines.

Effective grassroots organizations excel at attracting and retaining volunteers with education and/or experience in areas relevant to organizational mission and ongoing projects. The availability of intellectual capacity and talent in the community will impact the ability to successfully complete all but the most basic projects. As professionals with environmental expertise reach retirement age, the pool of potential volunteer expertise increases. As our cases illustrate, many of these people have both the time and the desire to apply their expertise to problems in their own communities, and this expertise can be employed to combine social capital, increased stewardship, and a commitment to place.[1] Related to this is the importance of competent, qualified staff. It takes the institutional core that the staff provides to transition from a person-based organization, dependent on the active involvement of the founders, to an ongoing institution with succession guidelines.

New grassroots organizations must attract the attention of the community. The use of symbolic, easily understood goals can grab the public's attention and enhance public excitement and involvement. LRN provides a perfect example of this with its original goal of being able to eat oysters from the river by the year 2007. Since oysters are very sensitive to pollution, the ability to harvest and eat oysters represents a very clean river and provides a goal that is more likely to rally support than one that involves a target level for nitrogen levels in the water. Eating oysters is a goal to which everyone in the community can relate.

It is also critical for new grassroots collaboratives to achieve at least some early successes. These early successes, no matter how small, will generate enthusiasm and commitment from volunteers and other stakeholders. Organizations should go for the "low-hanging fruit" first in order to cement the collaborative effort. This is a strategy that has worked for the organizations presented in our case studies. They can all point to some early successes that make it much easier to attract new volunteers and resources. It is human nature to want to be part of something that is seen as both important and successful. These successes not only develop legitimacy, but also efficacy—the collaborative builds a reputation as an organization capable of success, which in turn increases social capital.

Commitment to Place

The commitment to place, and its associated motivation, BIMBY (Because It's My Back Yard), is also evident in all three case studies. Place is very much a "local" phenomenon; in other words, the commitment to place almost always has a local focus which may help to explain why BIMBY is such a positive force for change or preservation. Many of our interviewees cited a personal connection to their watershed as their motivation for participation in the collaborative efforts. The "place" is not just the river itself and the people who live along its banks, but the community located within the watershed. Indeed, the industrial nature of the Elizabeth River means that very few people own waterfront property; most of those who do own waterfront land are the businesses located on the southern and eastern branches of the river. Nonetheless, people in the watershed have developed a connection to the river, and the efforts of the ERP to raise awareness of the threats to the river have galvanized additional efforts and programs. In the two cases that currently offer stewardship recognition programs to homeowners (River Star Homes and Pearl Homes), it is notable that most of the members of these programs are not waterfront residents. The key to utilizing this commitment to place appears to be to appeal to residents' sense of community, and then relying on members' willingness to convince their neighbors of the benefits of action. The commitment to place reflected in BIMBY may prove to be a more sustainable motivating force than NIMBY. "Because It's My Backyard" works as a motivating force even when there is no identifiable crisis, just degradation over time.

High-profile champions play a very significant role in the success of grassroots collaborative organizations. They act as central convener figures who are able to attract participants and have the respect of the stakeholder community. Every effort should be made to enlist champions and sponsors who are seen as influential members of the community. The likelihood of success in organizing grassroots environmental efforts is enhanced when stakeholders in the community are willing to listen and participate, and this is often dependent on the level of respect for the organization's founders. These conveners thus exercise significant social capital; since social capital is strongest at the community level, commitment to place and social capital can be thought of as two sides of the same coin.

Importance of Context

The context in which the collaborative activity takes place is also a critical concept in any understanding of grassroots, citizen-based collaboration. In short, context matters, and contexts differ. The identified problems in a watershed are likely to differ, as they do across our three cases. For example, the Elizabeth River is primarily industrial and polluted through point source pollution, whereas the Lynnhaven and Nansemond Rivers are primarily residential and the pollution to the rivers is almost all nonpoint source runoff. The nature of the rivers and the problems they face determines who the stakeholders are who must be included in restoration efforts. The breadth and diversity of stakeholders will influence the goals that grassroots collaboratives are able to pursue to accomplish their missions.

Likewise, the resources in the community also differ, as does the mix of stakeholders present in the community. Environmental issues are connected to the economic vitality of a community. There is a clear need to take into account the economic nature of the watershed for environmental collaboration to be successful. Political culture plays a role here as well, in that it helps determine the kinds of actions possible in the setting. While this research did not specifically explore variables such as per capita income, population density, education levels, racial and ethnic proportions, and so on, other research has found that these variables are also important indicators of the kinds of collaborative actions possible in a given area. Our research underscores the importance of context, and we encourage other scholars of collaboration more fully to consider the effects of contextual variables in their work.

Partnership with Stakeholders

This book also highlights the critical role of stakeholders, including government, in citizen-based collaborations. While each of our three case study groups was founded by citizens, each group regularly interacts with government agencies and policy makers at all three levels—federal, state, and local—as well as other stakeholders. Successful grassroots collaboratives are willing to collaborate with all stakeholders, not just selected ones. In the case of LRN, the ties between the citizen group and local government are especially important, and a strong synergistic relationship has developed between the two. The ERP historically has tended to focus more on relationships with state and federal agencies, although their attention has recently turned more toward local governments. The NRPA is new enough that it has little interaction with government, but there is a strong desire to develop closer ties with local government. Although the NRPA faces some rather daunting cultural and political barriers to overcome, the organization appears to be headed down a path that will likely result in the kinds of relationships it desires. Essentially, grassroots organizations can act as "force multipliers" for city governments and allow them to move environmental projects forward that they could not accomplish on their own. It should be recognized that citizen groups can do some things better than city governments (education and awareness) and vice versa (undertaking major resource-intensive projects and regulation). Additionally, nonprofit organizations should take the lead on efforts to change the behavior of citizens since nonprofits are almost always trusted by citizens more than government agencies.

The ties with government that do exist for all three organizations have led to increased legitimacy for both the organizations and their partner agencies. Partnerships with these environmental organizations have allowed governments to take actions that would otherwise be met with considerable public resistance; approval and support by these citizen groups lessens public criticism. Likewise, pressure to bring public resources to bear to meet environmental challenges, and success in doing so, has raised the visibility and "clout" of these groups, and leads observers to conclude that these groups are not "fly-by-night," but rather groups who possess important attributes. This cross-sectoral benefit also translates to private-sector organizations; in the case of the Elizabeth River, the ability of the ERP to bring businesses historically opposed to environmental action into a collaborative framework has significantly raised the standing of the ERP in the community.

Local Grassroots Environmental Collaborations Can *Make a Difference*

Finally, a more complicated approach to Bay restoration efforts may well be emerging—one that requires an appreciation of the possible limits of regional collaboration as well as the potential benefits of grassroots collaboration on a local level to complement "top down" requirements. Our research indicates that the ability of local grassroots collaborations to make a difference in water quality in their watersheds can indeed be successful, and that measurable changes in water quality can result from their efforts. These groups are able to mobilize stakeholders in their communities to focus on watershed issues, and provide a forum in which the stakeholders can not only participate in the actions of the collaborative, but participate in the goal-setting process as well. This level of participation allows participants to direct attention and resources to the goals of the group, thus enhancing the chances that positive outputs and outcomes will occur.

What is also clear is that these groups do not themselves cause environmental change; rather, they can empower stakeholders and participants in the group to direct resources toward the problem. In this regard, we conclude that many of the gains in environmental quality are the result of the "but for" effect—these outputs and outcomes would not occur "but for" the actions of the collaborative effort. LRN does not have the millions of dollars needed to address stormwater needs in the Lynnhaven watershed, but by bringing in the City of Virginia Beach as an important stakeholder, the collaborative group can provide both incentive and justification for the city to commit its resources to stormwater infrastructure upgrades. The city is willing to do this in part because it is the right thing to do, but also because its collaborative partners are fellow citizens in the community. The civic pride exhibited by these efforts, particularly in the *Bowling Alone* (Putnam 2000) era, gives city policy makers a clear measure of citizen support for their efforts. Moreover, it provides department-level personnel with a ready-made support mechanism for the department's programs. If the Department of Public Works builds stormwater sewers, to collaborate with a group interested in (and in need of) these resources creates a synergistic and interdependent relationship between the environmental group and the city. City employees often live in the same community, so they may also be influenced by the BIMBY effect. This process of integrating stakeholders with resources into a collaborative framework is the mechanism that allows collaboration to work. We contend that it is easier to entice stakeholders at the local level to contribute meaningfully to collaboration than it is at a larger scale.

However, we also note that local grassroots environmental collaborations have limits. Although this research does not identify the practical limits of "local collaboration," we suspect there is an upper limit to the size of the area (or perhaps the size of the population) in which collaboration will be successful. As the size of the collaboration grows, more stakeholders (and thus interests) must be incorporated. As the diversity of interests grows, more interests must be addressed (and satisfied) within the collaborative setting, which in turn complicates the goal-setting process. In addition, because levels of social capital are likely to be lower in larger groups, it is easier for some participants to adopt an adversarial approach to the process, which is antithetical to collaboration.

Collaboration with local government (as opposed to regional organizations and/or state government) can actually enhance command-and-control regulatory regimes, particularly those such as the new Bay TMDL requirements. By mobilizing citizen groups to champion best management practices (BMPs), many of which target unregulated nonpoint sources, local governments will be better able to meet the TMDL requirements for water-quality improvements in their local watersheds. In a very real way, local grassroots groups make success of the regulatory framework possible. As attention in the regulatory setting focuses increasingly on nonpoint pollution, behavior modification becomes exponentially more critical. The education efforts of these groups aimed at citizens can be much more effective than penalties tied to regulatory frameworks. Coupled with a national political setting that is increasingly hostile to the regulation of economic activity, new inducements must be emplaced to incentivize desired behaviors. The successes of our case-study groups suggest that inducements such as social capital and a commitment to place can succeed where more heavy-handed government regulatory frameworks cannot. In a very real way, these groups, and their focus on collaboration at the local level, can provide local governments with powerful allies and enrich the effectiveness of regulatory requirements.

Finally, one of the most important factors in the successful organization and management of the case-study collaboratives is the ability of the participants to "check their egos at the door." Grassroots collaboratives can conduct their business with a minimal amount of organizational dysfunction when participants are open-minded and welcome ideas and input from all participants and stakeholders alike. Organizational norms that emphasize the importance of civility in interactions will contribute to the satisfaction experienced by volunteers, staff, and stakeholders.

In sum, Bay-restoration efforts appear to be moving toward an era where local action by grassroots collaboratives will become increasingly important. Much of nonpoint source pollution caused by various and diffuse sources can only be controlled by individual behavior change—a far different challenge than presented by discrete point sources, which have been much more successfully controlled under the Clean Water Act's command-and-control scheme. Meeting many TMDLs requires a significant amount of voluntary action by individuals. Consequently, getting results through the TMDL process will be profoundly difficult—perhaps impossible without local grassroots support.

LOOKING TO THE FUTURE

This book is a small contribution to what should be a growing body of literature on grassroots collaborations. Grassroots collaborations differ not only in form, but also in efficacy, legitimacy, and effectiveness, from large-scale or agency-based collaborations. Collaboration as a means to address water-quality problems is not inherently good or bad; what matters is the scale at which the model is applied. Indeed, we believe that voluntary, citizen-based collaboration as a means to address water quality in a local setting is likely to result in more favorable outcomes. Likewise, command-and-control regulatory regimes are an important tool, but they have proven ineffective on their own, particularly when applied to problems such as nonpoint pollution. Our research suggests that a combination of regulatory frameworks and citizen-based collaboration applied at a local scale may offer the best chance to address the wicked problem of water pollution.

Our study should not be construed as a refutation of other approaches to water quality, nor should it be considered as the definitive "proof" of the efficacy of grassroots approaches to water-quality problems. Indeed, the limited context encompassing our three case studies is not ubiquitous enough to be generalizable to all other settings. We encourage others to apply our model of collaboration to other watersheds in other locations in an effort to confirm or refute our findings. Since our model (and our expectations) are drawn from a growing literature on collaboration, we expect our findings to be repeatable in a range of other settings. However, at this point the utility of our model is very much an empirical question.

An important element missing from our work is a comparative case of an unsuccessful grassroots collaborative organization. Of the seven organizations available to us for in-depth qualitative study in our region, none could reasonably be said to be "unsuccessful." Several, including our cases, have made missteps along the way, but all have learned from those mistakes. More importantly, none have experienced mistakes due to their commitment to authentic collaboration.

Regardless of the ultimate status of our model, we do believe that the lessons drawn from our cases can help inform other organizations and individuals engaged in similar efforts. These lessons are consistent with the expectations found in the literature; they thus give us confidence that our findings are repeatable, but they also give us confidence to recommend those lessons to others. We make no claim that all lessons will work in all situations, conditions, or settings, but each is worthy of consideration for implementation elsewhere. We encourage other collaborative organizations to review these points, and to apply those that seem relevant and useful.

The problem of how best to address water quality remains a "wicked problem." Local-level grassroots collaboration relies very much on BIMBY as a motivation for action, and can serve as a focal point for successful water quality action. To paraphrase Captain Bob Crocker (2012) of the Volvo Penta Testing Facility in Suffolk, the best way to clean up the water may be one community at a time.

NOTE

1. We must point out that Hampton Roads may be somewhat unique in this regard, given the significant presence of national, state, and regional organizations with a presence in the region. However, the transferrable "lesson" is that retired persons possess both time and expertise, and can greatly enhance the capabilities of grassroots organizations.

Appendix

This appendix lists the individuals interviewed for our study, and their affiliation.
Key:

ERP	Elizabeth River Project
LRN	Lynnhaven River NOW
NRPA	Nansemond River Preservation Alliance

All others listed by their current institutional affiliation

Last Name	First Name	Organization
Adams	Sharon	ERP
Barnett	Cheryl	ERP
Barnum	Steven	NRPA
Barrow	Mary Reid	LRN
Barry	Richard	NRPA
Bernick	Clay	City of Virginia Beach
Boatwright	Pam	ERP
Bradley	Bruce	ERP
Bray	Bobby	ERP (Virginia Port Authority)
Chalmers	Hap	LRN (Lynnhaven Oyster Company)
Copeland	Howard	ERP
Corbin	Jeff	US Environmental Protection Agency
Crocker	Bob	NRPA (Volvo Penta)
Daniel	Frank	ERP
Davis-Martin	James	VA Department of Conservation and Recreation
Dean	Robert	ERP

DuBois	Kevin	LRN
Dunbar	Robin	ERP
Eure	John	NRPA
Everett	Christy	Chesapeake Bay Foundation
Ewing	Mike	ERP (BAE Systems)
Fine	Andrew	LRN
Forget	Karen	LRN
Frankenfield	Barry	City of Virginia Beach
Gates	Janice	NRPA; ERP; LRN
Gonzalez	Gretchen	NRPA
Harper	Scott	*Virginian-Pilot*
Heinricht	Mary	ERP
Henifin	Ted	Hampton Roads Sanitation District
Herbert	Steven	City of Virginia Beach
Hinshelwood	Geoffrey	NRPA
Host	Mike	ERP (Norfolk Naval Shipyard)
Hume	Bob	ERP
Jackson	Fleta	City of Norfolk
Jackson	Marjorie M.	ERP
Johnston	Robert	City of Virginia Beach
Katchmark	Whitney	Hampton Roads Planning District Commission
Kensler	Mike	ERP
Koubsky	Dave	ERP
Lester	Harry	LRN
Linkenhoker	Cindy	Parsons-Brinckerhoff
Lott	Nathan	Virginia Conservation Network
Martin	Howard	NRPA
Marx	Peter	Choose Clean Water Coalition
McCarthy	Jerry	Virginia Environmental Endowment
Moses	Ray	ERP
Neikirk	Chris	ERP (Norfolk Southern)
Nickelsburg	Mike	ERP

Oliver	Jim	Hampton Roads Center for Civic Engagement
Pawlukiewicz	Janet	US Environmental Protection Agency (retired)
Peace	Whit	LRN
Picardi	Mary	LRN
Priest	Josh	ERP
Priest	Walter	National Oceanographic and Atmospheric Administration
Psimas	Elizabeth	ERP (City of Portsmouth)
Reisch	Tim	ERP
Rieger	Joe	ERP
Ritter	Debbie	ERP (City of Chesapeake)
Ritter	Ron	ERP
Rule	Joe	ERP
Serpe	Richard	LRN
Smith	Karla	NRPA
Sorabella	Laurie	LRN (Oyster Reef Keepers of VA)
Spore	James	City of Virginia Beach
Stanton	Robert	LRN
Stedfast	Phil	ERP (Kinder Morgan)
Stewart	John	Lafayette Wetlands Partnership
Stiles	Skip	Wetlands Watch
Strutton	Randi	Hoffler Creek Wildlife Foundation
Taraski	Elizabeth	NRPA
Thomas	Joe	ERP
Tilton	Mary	Back Bay Restoration Foundation
Tribo	Jenny	Hampton Roads Planning District Commission
Unger	Mike	Virginia Institute of Marine Science/ERP
Wass	John	NRPA
Wayland	Bob	US Environmental Protection Agency (retired)
Wood	Heather	Virginia Port Authority

Worth	Nicholas	NRPA
Wyatt	Dottie	ERP

References

Adger, W. Neil. 2003. Social capital, collective action, and adaptation to climate change. *Economic Geography* 79:387–404.

Adams, Sharon. 2012a. Interview with William Gibson. Tape recording, January 13.

———. 2012b. Interview with William Gibson. Tape recording, March 1.

Adler, Robert W., Jessica C. Landman, and Diane M. Cameron. 1993. *The Clean Water Act twenty years later*. Washington, DC: Island Press.

Agranoff, Robert, and Michael McGuire. 2001. Big questions in public network management research. *Journal of Public Administration Research and Theory* 11:295–326.

Altheide, David L., and John M. Johnson. 1998. Criteria for assessing interpretive validity in qualitative research. In *Collecting and Interpreting Qualitative Materials*, ed. Norman K. Denzin and Yvonna S. Lincoln, 283–312. Thousand Oaks, CA: Sage.

APM Terminals. 2012. *Terminal information*. www.apmterminals.com/americas/virginia/terminalinfo.aspx.

Barnett, Cheryl. 2012. Interview with William Gibson. Tape recording, July 13.

Barnum, Steven. 2012. Interview with John C. Morris. Tape recording, May 31.

Barrow, Mary Reid. 2012. Interview with William M. Leavitt. Tape recording, November 12.

_____. 2004. Saving the Lynnhaven oysters: From plight to plate. *Virginian-Pilot* (Norfolk, VA), February 29.

Barry, Richard. 2012. Interview with John C. Morris. Tape recording, May 24.

Bernick, Clay. 2012. Interview with William M. Leavitt. Tape recording, April 18.

Bidwell, Ryan D., and Clare M. Ryan. 2006. Collaborative partnership design: The implications of organizational affiliation for watershed partnerships. *Society and Natural Resources* 19:827–43.

Bingham, Lisa Blomgren, Rosemary O'Leary, and Christine Carlson. 2008. Frameshifting: Lateral thinking for collaborative public management. In *Big Ideas in Collaborative Public Management*, ed. Lisa Blomgren Bingham and Rosemary O'Leary, 3–16. Armonk, NY: M.E. Sharpe.

Blackistone, Mick. 2001. *Dancing with the tide: Watermen of the Chesapeake*. Centreville, MD: Tidewater Publishers.

Blankenship, Karl. 2008. Proof or consequences: Latest cleanup plans must meet goals. *Bay Journal*, October 18. www.bayjournal.com/article/proof_or_consequences_latest_cleanup_plans_must_meet_goals.

Boatwright, Pamela. 2012a. Interview with William Gibson. Tape recording, June 15.

_____. 2012b. Telephone interview with William Gibson. December 14.

Boesch, Donald F., Russell B. Brinsfield, and Robert E. Magnien. 2001. Chesapeake Bay eutrophication: Scientific understanding, ecosystem restoration, and challenges for agriculture. *Journal of Environmental Quality* 30:303–20.

Bonnell, Joseph E., and Tomas M. Koontz. 2007. Stumbling forward: The organizational challenges of building and sustaining collaborative watershed management. *Society and Natural Resources* 20:153–67.

Bourdieu, Pierre. 1985. The forms of capital. In *Handbook of Theory and Research for the Sociology of Education*, ed. John G. Richardson, 241–58. New York: Greenwood Press.

Bradley, Bruce. 2012. Interview with William Gibson. Tape recording, April 3.

Bray, Robert. 2012. Interview with William Gibson. Tape recording, October 5.

Bryson, John, Barbara Crosby, and Melissa M. Stone. 2006. The design and implementation of cross-sector collaborations: Propositions from the literature. *Public Administration Review* 66:44–53.

Campbell, Joseph T., Tomas M. Koontz, and Joseph E. Bonnell. 2011. Does collaboration promote grass-roots behavior change? Farmer adoption of best management practices in two watersheds. *Society and Natural Resources* 24:1127–41.

Carmin, JoAnn, Nicole Darnall, and Joao Mil-Homens. 2003. Stakeholder involvement in the design of U.S. voluntary environmental programs: Does sponsorship matter? *Policy Studies Journal* 31:527–43.

Carson, Rachel. 1962. *Silent Spring*. Boston: Houghton Mifflin.

Caulfield, Henry P. 1989. The conservation and environmental movements: An historical analysis. In *Environmental Politics and Policy: Theories and Evidence*, ed. James P. Lester, 13–56. Durham, NC: Duke University Press.

Chalmers, Hap. 2012. Interview with William M. Leavitt. Tape recording, October 3.

Chesapeake Bay Agreement 2000. www.chesapeakebay.net/content/publications/cbp_12081.pdf.

_____. 1987. www.chesapeakebay.net/content/publications/cbp_12510.pdf.

_____. 1983. www.chesapeakebay.net/content/publications/cbp_12512.pdf.

Chesapeake Bay Agreement Amendments. 1992. www.chesapeakebay.net/content/publications/cbp_12507.pdf.

Chesapeake Bay Commission. 2003. *The Cost of a Clean Bay: Assessing Funding Needs Throughout the Watershed*. Annapolis, MD: Author.

Chesapeake Bay Foundation. 2012a. About CBF: History. www.cbf.org/about-cbf/history.

_____. 2012b. *Saving a National Treasure*. www.cbf.org/how-we-save-the-bay/programs-initiatives/hampton-roads-virginia/lafayette-river-restoration.

_____. 2010. Lynnhaven River points the way to a clean bay. *Bay Daily*, October 22. cbf.typepad.com/bay_daily/2010/10/lynnhaven-river-points-the-way-to-a-clean-bay.html.

_____. 2009. *The Economic Argument for Cleaning Up the Chesapeake Bay and its Rivers*. www.cbf.org/document.doc?id=1094.

_____. 2006. Restoration sites in south Hampton Roads. www.chesbay.us/Presentations/November2006/Priest.pdf.

_____. 2005. *Lower James River Fact Sheet*. www.cbf.org/document.doc?id=233.

Chesapeake Bay Program. 2012a. Discover the Bay watershed. www.chesapeakebay.net/discover/baywatershed.

_____. 2012b. Facts & figures. www.chesapeakebay.net/discover/bay101/facts.

_____. 2012c. Chesapeake Bay Stat. stat.chesapeakebay.net/?q=node/130&quicktabs_10=0.

Choose Clean Water Coalition. 2012. About us. choosecleanwater.org/the-coalition/about-us/.

Cigler, Beverly A. 1999. Pre-conditions for the emergence of multicommunity collaborative organizations. *Policy Studies Review* 16:87–102.

City of Suffolk. 2012. All about Suffolk: History. www.suffolkva.us/community/history.html.

Clancy, Paul. 2011. Fans of river with a rich past seek a robust future. *Virginian-Pilot* (Norfolk, VA), April 24.

Clark, Brad T., Nina Burkardt, and M. Dawn King. 2005. Environmental assessment watershed management and organizational dynamics: Nationwide findings and regional variation. *Environmental Management* 36:297–310.

Clean Waters Task Force. 2012. Minutes, April 12. City of Virginia Beach, VA.

Cline, Sarah, and Alan Collins. 2002. Watershed associations in West Virginia: Their impact on environment protection. *Journal of Environmental Management* 67:373–82.

Coccaro, Joseph. 1993. Group forms to tackle Elizabeth River: pollution members are unsure of how best to help the fouled waterway. *Virginian-Pilot* (Norfolk, VA), June 10.

Coleman, James S. 1988. Social capital in the creation of human capital. *American Journal of Sociology* 94:S95–S120.

College of Business and Public Administration. 2010. *Hampton Roads versus other east coast container ports.* bpa.odu.edu/forecasting/sor/2010/2010_SOR_Container_Ports.pdf.

Congressional Budget Office (CBO). 1994. The total costs of cleaning up nonfederal Superfund sites. Washington, DC: Author.

Conley, Alexander, and Margaret Moote. 2003. Evaluating collaborative natural resource management. *Society and Natural Resources* 16:371–86.

Cooper, Terry, Thomas Bryer, and Jack Meek. 2006. Citizen-centered collaborative public management. *Public Administration Review* 66 (Special Issue, December): 76–88.

Copeland, Howard. 2012. Interview with William Gibson. Tape recording, July 5.

Corbin, Jeff. 2012. Interview with Shana Jones. Tape recording, July 10.

Crisman, Phoebe, and Michael Petrus. 2007. *The Money Point Sustainable Revitalization Plan—Chesapeake, VA.* University of Virginia: School of Architecture. www.arch.virginia.edu/learningbarge/Community%20History%20Website/Website%20Edit/Money_Point/mny_pt_rvtlztn_plan.pdf.

Crocker, Robert. 2012. Interview with John C. Morris. Tape recording, May 25.

Culhane, Paul J. 1981. *Public Lands Politics: Interest Group Influence on the Forest Service and the Bureau of Land Management.* Baltimore, MD: Johns Hopkins Univerisity Press.

Curtis, Allan, Bruce Shindler, and Angela Wright. 2002. Sustaining local watershed initiatives: Lessons from landcare and watershed councils. *Journal of the American Water Resources Association* 38:1207–16.

Dabney, Virginius. 1971. *Virginia: The New Dominion, a History from 1607 to the Present.* Garden City, NY: Doubleday.

Daley, Dorothy. 2008. Interdisciplinary problems and agency boundaries: Exploring effective cross-agency collaboration. *Journal of Public Administration Research and Theory* 19:477–93.

Daniel, Frank. 2012. Interview with William Gibson. Tape recording, March 15.

Darnall, Nicole, and JoAnn Carmin. 2005. Greener and cleaner? The signaling accuracy of U.S. voluntary environmental programs. *Policy Science* 38:71–90.

Darnall, Nicole, Joann Carmin, Nicole Kreiser, and Joao Mil-Homens. 2003. *The Design and Rigor of U.S. Voluntary Environmental Programs: Results from the VEP Survey.* Department of Policital Science & Public Administration, North Carolina State University and Department of Urban Studies and Planning, Massachusetts Institute of Technology.

Darnall, Nicole, and Stephen Sides. 2008. Assessing the performance of voluntary environmental programs: Do certifications matter? *Policy Studies Journal* 36:95–117.

Davis-Martin, James. 2012. Interview with Shana Jones. Tape recording, July 10.

Davison, Steven, Jay Merwin, Jr., John Capper, Garrett Power, and Frank Shivers, Jr. 1997. *Chesapeake Waters: Four Centuries of Controversy, Concern, and Legislation,* 2nd ed. Centreville, MD: Tidewater Publishers.

Delmas, Magali A., and Ann K. Terlaak. 2001. A framework for analyzing enviromental voluntary agreements. *California Management Review* 43:44–63.

DuBois, Kevin. 2012. Interview with William M. Leavitt. Tape recording, April 16.

Dunbar, Robin. 2012. Interview with William Gibson. Tape recording, October 15.

Duram, Leslie A., and Katharin G. Brown. 1999. Assessing public participation in US watershed planning. *Society and Natural Resources* 12:455–67.

Easton, David. 1971. *The Political System: An Inquiry into the State of Political Science,* 2nd ed. New York: Alfred A. Knopf.

_____. 1965. *A Framework for Political Analysis.* Englewood Cliffs, NJ: Prentice Hall.

Elazar, Daniel J. 1984: *American Federalism: A View from the States,* 3rd ed. New York: Harper and Row.

Elizabeth River Project. 2012a. Superfund sites. www.elizabethriver.org/the_elizabeth_river/superfund_sites.aspx.

_____. 2012b. Congratulations River Star businesses 2012. www.elizabethriver.org/PDFs/RiverStarIndustries/2012%20River%20Star%20Businesses%20List.pdf.

_____. 2012c. River Star businesses. www.elizabethriver.org/RiverStars/RiverStarsIndustires.aspx.

_____. 2012d. The Elizabeth River Project: Unrestricted vs. restricted budget funding-draft. December 23.

_____. 2012e. Money Point: Fish and shellfish approve of restoration work. www.elizabethriver.org/Projects/Money_Point.aspx.

_____. 2012f. Students and teachers rave about the 2012 Learning Barge field trips. www.elizabethriver.org/Projects/The_Learning_Barge.aspx.

_____. 2012g. River Star homes. www.elizabethriver.org/RiverStars/RSH-Sign-up_Form.aspx.

_____. 2012h. Seven easy steps to become a River Star home. *Mudflats*. www.elizabethriver.org/PDFs/Mudflats/MudFlats_0920b.pdf.

_____. 2011a. *Chesapeake Bay Foundation and Elizabeth River Project—The Plan for Restoring the Lafayette River: Strategies for Community-wide Action: Executive Summary*.

_____. 2011b. *The Plan for Restoring the Layayette River.* April 27. www.elizabethriver.org/PDFs/Lafayette%20River%20Restoration/Lafayette%20River%20Restoration%20Plan.pdf.

_____. 2009. Elizabeth River "goo is going" set for July 1. www.elizabethriver.org/PressRoom/June-29-Money-Point-Press-Release.pdf.

_____. 2008a. *State of the River Report*. Portsmouth, VA: Author.

_____. 2008b. *One Creek at a Time.* June 30. www.elizabethriver.org/PDFs/Paradise-Creek/Paradise_Creek_Case_Study-FINAL-08-12-08.pdf.

_____. 2008c. *The River of the Future: A Watershed Action Plan*, revised 3rd ed. Portsmouth, VA: Author.

_____. 2008d. *Balancing Industry and the Environment: How to Achieve Win-Win on the Industrial Waterfront*. Portsmouth, VA: Author.

_____. 2006a. *Rediscover the Treasure: Money Point Revitalization.* www.elizabethriver.org/PDFs/MoneyPoint/MP_layout_061011final.pdf.

_____. 2006b. *Money Point. Almost $15 Million Underway!* www.elizabethriver.org/PDFs/MoneyPoint/MoneyStory_MoneyPoint_proj2010-19-6.pdf.

_____. 2003. *Paradise Found: Paradise Creek Restoration Plan.* August 1. www.virginia.edu/ien/moneypoint/mpointfiles/Paradise%20Found.pdf.

_____. 2002. *Charter for the Sedimentation Remediation Partnership Committee.* Portsmouth, VA: Author.

_____. 1998. *P2 Workbook.* www.elizabethriver.org/Publications/ERP_Pubs/Pollution%20Prevention%20Workbook.pdf.

_____. 1996. *Elizabeth River Restoration: A Watershed Action Plan.* June 20. Portsmouth, VA: Author.

_____. 1995. *Elizabeth River Dimensions: Assessment and Ranking of Risks to the Ecosystem, Human Health and Quality of Life*. Portsmouth, VA: Author.

_____. 1993. *Elizabeth River Visions.* October 22. Portsmouth, VA: Author.

_____. 1992. *Elizabeth River Project Report: Initial Phase.* Portsmouth, VA: Author.

Elizabeth River Restoration Trust Operating Agreement. 2004. www.deq.virginia.gov/Portals/0/DEQ/Water/WetlandsStreams/eliztrustoperatingagree.pdf.

Emerson, Kirk, Tina Nabatchi, and Stephen Balogh. 2012. An integrative framework for collaborative governance. *Journal of Public Administration Research and Theory* 22:1–29.

Ernst, Howard R. 2010. *Fight for the Bay: Why a Dark Green Environmental Awakening Is Needed to Save the Chesapeake Bay*. Lanham, MD: Rowman & Littlefield.

_____. 2003. *Chesapeake Bay Blues: Science, Politics, and the Struggle to Save the Bay.* Lanham, MD: Rowman & Littlefield.

Eure, John. 2012. Interview with John C. Morris. Tape recording, May 17.

Everett, Christy. 2012. Interview with Shana Jones. Tape recording, February 28.
Ewing, Michael. 2012a. Interview with William Gibson. Tape recording, June 19.
_____. 2012b. Interview with William Gibson. Tape recording, October 10.
Executive Order No. 13,508. Chesapeake Bay protection and restoration. 2010.
Federal Register. 2011. Notice for the establishment of the total maximum daily load (TMDL) for the Chesapeake Bay, 76 Fed. Reg. 549.
_____. 2009. EPA issues notice about Clean Water Act, September 18.
Fine, Andrew. 2012. Interview with William M. Leavitt. Tape recording, March 23.
Flint, Courtney G., A. E. Luloff, and James C. Finley. 2008. Where is "community" in community-based forestry? *Society and Natural Resources* 21:526–37.
Floress, Kristin, J. Mangun, M. Davenport, and K. Williard. 2009. Constraints to watershed planning: Group structure and process. *Journal of The American Water Resources Association* 45:1352–60.
Floress, Kristin, Linda S. Prokopy, and Shorna B. Allred. 2011. It's who you know: Social capital, social networks and watershed groups. *Society and Natural Resources* 24:871–86.
Forget, Karen. 2012a. Interview with William M. Leavitt. Tape recording, February 23.
_____. 2012b. Interview with William M. Leavitt. Tape recording, June 25.
_____. 2012c. Interview with William M. Leavitt. Tape recording, February 2.
_____. 2012d. Interview with William M. Leavitt. Tape recording, April 10.
_____. 2011. Introduction. In *The Lynnhaven: Restoring a Legend*. Virginia Beach, VA: Lynnhaven River NOW.
Frankenberry, Rita. 2011. Archaeologists hunt for traces of colonial-era Kempe's Landing. *Virginian-Pilot/HamptonRoads.com*. August 7. hamptonroads.com/2011/08/archaeologists-hunt-traces-colonialera-kempes-landing.
Frankenfield, Barry. 2012. Interview with William M. Leavitt. Tape recording, April 24.
Gates, Janice. 2012. Interview with John C. Morris. Tape recording, June 28.
Geertz, Clifford. 1973. Thick description: Toward an interpretive theory of culture. In *The Interpretation of Cultures: Selected Essays*, ed. Clifford Geertz, 3–30. New York: Basic Books.
Genskow, Kenneth. 2009. Catalyzing collaboration; Wisconsin's agency-initated basin partnerships. *Environmental Management* 43:411–24.
Genskow, Kenneth D., and Stephen M. Born. 2006. Organizational dynamics of watershed partnerships: A key to integrated water resources managment. *Journal of Contemporary Water Research & Education* 35:56–64.
Gerlak, Andrea K. 2005. Federalism and U.S. water policy: Lessons for the twenty-first century. *The Journal of Federalism* 36:231–57.
Gerlak, Andrea, and Tanya Heikkila. 2007. Collaboration and institutional endurance in U.S. Water Policy. *Political Science and Politics* 40:55–60.
Goldfarb, William. 1994. Watershed management: Slogan or solution? *Boston College Environmental Affairs Law Review* 21:483–510.
Gonzalez, Gretchen. 2012. Interview with John C. Morris. Tape recording, May 17.
Gray, Barbara. 1989. *Collaborating: Finding Common Ground f or Multiparty Problems*. San Francisco: Jossey-Bass.
———. 1985. Conditons facilitating interorganizational collaboration. *Human Relations* 38:911–36.
Groseclose, Chase. Oysterman's return. In *The Lynnhaven: Restoring a Legend*, 55. Virginia Beach, VA: Lynnhaven River NOW.
Hampton Roads Planning District Commission. 2012a. About us. www.hrpdcva.gov/page/about.
_____. 2012b. *Chesapeake Bay Phase II Watershed Implementation Plan: Hampton Roads Regional Planning Framework, Scenario, and Strategies*. Report No. PEP-12-01. www.hrpdcva.gov/Documents/Phys%20Planning/2012/ChesBayTMDL/PEP-12-01_HR_PhIIWIP_RegionalAppendix.pdf.
_____. 2011. *Cost Estimates for the Chesapeake Bay TMDL*. Chesapeake, VA: Author.

Hardy, Scott. 2010. Governments, groups membership, and watershed partnerships. *Society and Natural Resources* 23:587–603.
Hardy, Scott D., and Tomas Koontz. 2009. Rules for collaboration: Institutional analysis of group membership and levels of action in watershed partnership. *The Policy Studies Journal* 37:393–414.
Harper, Scott. 2012a. Elizabeth River Project wins $140k federal grant. *Virginian-Pilot* (Norfolk, VA), September 19.
_____. 2012b. Interview with Shana Jones. Tape recording, Jun. 6.
_____. 2012c. EPA withholds aid until state improves bay's pollution diet. *Virginian-Pilot* (Norfolk, VA), September 11.
_____. 2012d. Lynnhaven's friends mark progress, dig in for more. *Virginian-Pilot* (Norfolk, VA), January 27.
_____. 2012e. Grants for Lafayette aim to speed river restoration. *PilotOnLine.* hamptonroads.com/2012/04/grants-lafayette-aim-speed-river-restoration.
_____. 2011a. Leaders decide against suing EPA over bay "pollution diet." *Virginian-Pilot* (Norfolk, VA), April 1.
_____. 2011b. Activists unveil plan to reduce pollution in the Lafayette River. *Virginian-Pilot* (Norfolk, VA), April 28.
_____. 2011c. Cleanup project targets homeowners. *Daily News* (Newport News, VA), October 24.
_____. 2011d. River program in Norfolk aims to reduce pollutants. *Virginian-Pilot* (Norfolk, VA), October 7.
_____. 2011e. One river, many differences. *Virginian-Pilot* (Norfolk VA), June 19.
_____. 2010. Corps branch cited for its oyster work. *Virginian-Pilot* (Norfolk, VA), January 15.
_____. 2006. Green up top. *Virginian-Pilot* (Norfolk, VA), May 22.
_____. 2004. With details hashed out, river fund gets under way. *Virginian-Pilot* (Norfolk, VA), January 26.
_____. 2003. Maersk project clears key hurdle; additional millions for cleanup prompt EPA to support plan. *Virginian-Pilot* (Norfolk, VA), August 22.
_____. 1998. Money talks for Elizabeth River: Army corps of engineers to study ways to revive river. *Virginian-Pilot* (Norfolk, VA), June 6.
_____. 1995. Group explores ways to clean up Elizabeth River: Suggestions range from education to cap on construction. *Virginian-Pilot* (Norfolk, VA), January 11.
Healthy Harbor Case Studies. 2010. *The Elizabeth River.* www.healthyharborbaltimore.org/uploads/file/Healthy_Harbor_Case_Studies.pdf.
Heikkila, Tanya, and Andrea Gerlak. 2005. The formation of large-scale collaborative resource management institutions: Clarifying the roles of stakeholders, science, and institutions. *Policy Studies Journal* 33:583–612.
Heilman, John G., and Gerald W. Johnson. 1992. *The Politics and Economics of Privatization: The Case of Wastewater Treatment.* Tuscaloosa: University of Alabama Press.
Heinricht, Mary. 2012. Interview with William Gibson. Tape recording, March 22.
Henifin, Ted. 2012. Interview with Shana Jones. Tape recording, August 22.
Herbert, Steve. 2012. Interview with William M. Leavitt. Tape recording, May 4.
Herbert, Todd. 2012. Interview with Shana Jones. Tape recording, July 10.
Hill, Noah. 2012. Interview with Shana Jones. Tape recording, Jul. 10.
Hines, Michael. 2002. Protectors of the bay foundation marks 20 years in Virginia. *The Daily Press* (Newport News, VA), April 28.
Hinshelwood, Geoffrey. 2012. Interview with John C. Morris. Tape recording, July 25.
Horton, Tom, and William Eichbaum. 2003. *Turning the Tide: Saving the Chesapeake Bay.* Washington, DC: Island Press.
Host, Mike. 2012. Interview with William Gibson. Tape recording, June 21.
Houck, Oliver. 2011. The Clean Water Act returns (again): Part I, TMDLs and the Chesapeake Bay. *Environmental Law Reporter* 41:10208–228.
_____. 2002. *The Clean Water Act TMDL program: Law, policy, and implementation*, 2nd ed. Washington, DC: Environmental Law Institute.

Hume, Robert. 2012. Interview with William Gibson. Tape recording, June 27.

Imperial, Mark. 2005. Using collaboration as a governance strategy: Lessons from six watershed management programs. *Administration & Society* 37:281–320.

Irvin, Renee, and John Stansbury. 2004. Citizen participation in decision making: Is it worth the effort? *Public Administration Review* 64:55–64.

Jackson, Marjorie Mayfield. 2012a. Interview with William Gibson. Tape recording, March 6.

_____. 2012b. Interview with William Gibson. Tape recording, June 1.

_____. 2012c. Interview with William Gibson. Tape recording, December 14.

_____. 2006. Money for contaminated Money Point goes a long way. *Virginian-Pilot* (Norfolk, VA), November 6.

John, DeWitt. 1994. *Civic Environmentalism: Alternatives to Regulation in States and Communities.* Washington, DC: CQ Press.

Johnson, Janet B., and Richard A. Joslyn. 1991. *Political Science Research Methods,* 2nd ed. Washington, DC: CQ Press.

Johnston, Robert. 2012. Interview with William M. Leavitt. Tape recording, September 27.

Judy, Chris (Maryland Department of Natural Resources). 2012. How do oysters help improve water quality in the Bay? Available at dnr.md.gov/mydnr/askanexpert/oysters_water_qual.asp.

Kenney, Douglas. 2000. *Arguing about Consensus: Examining the Case against Western Watershed Initiatives and Other Collaborative Groups Active in Natural Resources Management.* Boulder: Natural Resources Law Center, University of Colorado School of Law.

Kenney, Douglas S., Sean T. McAllister, William H. Caile, and Jason S. Peckham. 2000. *The New Watershed Source Book.* Boulder: Natural Resources Law Center, University of Colorado School of Law.

Kensler, Mike. 2012. Interview with Shana Jones. Tape recording, March 23.

Kettl, Donald F. 2006. Managing boundaries in American administration: The collaboration imperative. *Public Administration Review 66* (Special Issue, December): 10–19.

_____. 2002. *The Transformation of Governance: Public Administration for Twenty-First Century America.* Baltimore: The Johns Hopkins University Press.

_____. 2000. The transformation of governance: Globalization, devolution, and the role of government. *Public Administration Review* 60:488–97.

Kingdon, John. 1984. *Agendas, Alternatives, and Public Policies*. Boston: Little, Brown.

Kobell, Rona. 2011. Elizabeth River rises from the depths. *Chesapeake Bay Journal* (Annapolis, MD), July–August.

Koehler, Dinah. 2007. The effectiveness of voluntary environmental programs: A policy at a crossroads? *The Policy Studies Journal* 35:689–722.

Koontz, Tomas, and Elizabeth Johnson. 2004. One size does not fit all: Matching breadth of stakeholder participation to watershed group accomplishments. *Policy Science* 37:185–204.

Koontz, Tomas, and Craig Thomas. 2006. What do we know and need to know about the environmental outcomes of collaborative managment? *Public Administration Review* 66 (Special Issue, December): 111–21.

Koubsky, David. 2012. Interview with William Gibson. July 5.

Kraft, Michael, and Denise Scheberle. 1998. Environmental federalism at decade's end: New approaches and strategies. *Publius* 28:131–46.

Kraft, Michael E., and Norman J. Vig. 1994. Environmental policy from the 1970s to the 1990s: Continuity and change. In *Environmental Policy in the 1990s,* 2nd ed., ed. Norman Vig and Michael E. Kraft, 3–30. Washington, DC: CQ Press.

Lake, Ronald La Due, and Huckfeldt, Robert. 1998. Social capital, social networks, and political participation. *Political Psychology* 19:567–84.

Leach, William D. 2006. Collaborative public management and democracy: Evidence from western watershed partnerships. *Public Administration Review 66* (Special Issue, December 2006): 100–10.

Leach, William, and Neil Pelkey. 2001. Making watershed partnerships work: A review of the empirical literature. *Journal of Water Resources Planning and Management* 127:378–85.

Leach, William, Neil Pelkey, and Paul Sabatier. 2002. Stakeholder partnerships as collaborative policymaking: Evaluation criteria applied to watershed management in California and Washington. *Journal of Policy Analysis and Management* 21:645–70.

Leach, William, and Paul Sabatier. 2005. Are trust and social capital the key to success? Watershed partnerships in California and Washington. In *Swimming Upstream: Collaborative Approaches to Watershed Management*, by Paul Sabatier, W. Focht, M. Lubell, Z. Trachtenberg, A. Vedlitz, and M. Matloack, 233–58. Cambridge, MA: MIT Press.

Leahy, Jessica, and Dorothy Anderson. 2010. "Cooperation gets it done": Social capital in natural resource management along the Kaskaskia River. *Society and Natural Resources* 23:224–39.

Lee, Robert, and Elen Stokes. 2009. Environmental governance: Reconnecting the global and local. *Journal of Law and Society* 36:1–10.

Lester, Harry. 2012. Interview with William M. Leavitt. Tape recording, March 28.

Lin, Nan. 2001. *Social Capital: A Theory of Social Structure and Action*. New York: Cambridge University Press.

Linkenhoker, Cindy. 2012. Interview with John C. Morris. Tape recording, June 13.

Lott, Nathan. 2012. Interview with Shana Jones. Tape recording, October 9.

Love, Cliff. 2011. Oyster gardening and beyond. In *The Lynnhaven: Restoring a Legend*. Virginia Beach, VA: Lynnhaven River NOW.

Lubell, Mark. 2004a. Collaborative watershed management: A view from the grassroots. *Policy Studies Journal* 32:341–61.

______. 2004b. Collaborative environmental institutions: All talk and no action? *Journal of Policy Analysis and Management* 23:549–73.

Lubell, Mark, Paul A. Sabatier, Arnold Vedilitz, Will Focht, Zev Trachtenberg, and Marty Matlock. 2005. Conclusions and Recommendations. In *Swimming Upstream: Collaborative Approaches to Watershed Management*, by Paul A. Sabatier, Will Focht, Mark Lubell, Zev Trachtenberg, Arnold Vedlitz, and Marty Matlock, 261–96. Cambridge, MA: MIT Press.

Lubell, Mark, Mark Schneider, John T. Scholz, and Mihriye Mete. 2002. Watershed partnerships and the emergence of collective action institutions. *American Journal of Political Science* 46:148–63.

Lurie, Susan, and Michael Hibbard. 2008. Community-based natural resource management: Ideals and realities for Oregon watershed councils. *Society and Natural Resources* 21:430–40.

Lynnhaven River NOW (LRN). 2012a. *Army Corps of Engineers oyster reef restoration*. www.lynnhavenrivernow.org/army-corps.aspx.

_____. 2012b. *Bylaws of Lynnhaven River NOW*. Virginia Beach, VA: Author.

_____. 2012c. Home page. www.lynnhavenrivernow.org/.

_____. 2011. 2011 *State of the River Report*. 2011. Virginia Beach, VA: Author.

_____. 2010. *2010 State of the River Report*. 2010. Virginia Beach, VA: Author.

Lyon, Thomas P., and John W. Maxwell. 2007. Environmental public voluntary programs reconsidered. *Policy Studies Journal* 36:723–50.

Maddry, Lawrence. 1996. Join in group's effort to restore the Elizabeth River. *Virginian-Pilot* (Norfolk, VA), June 19.

Mandarano, Lynn. 2008. Evaluating collaborative environmental planning outputs and outcomes: Restoring and proctecting habitat and the New York-New Jersey Harbor Estuary Program. *Journal of Planning Education and Research* 27:456–68.

Margerum, Richard D. 2008. A typology of collaboration efforts in environmental management. *Environmental Management* 41:487–500.

_____. 2007. Overcoming locally based collaboration constraints. *Society and Natural Resources* 20:135–52.

_____. 2002. Evaluating collaborative planning: Implications from an empirical analysis of growth management. *Journal of the American Planning Association* 68:179–93.

Martin, Howard. 2012. Interview with John C. Morris. Tape recording, May 21.

Marx, Peter. 2012. Interview with Shana Jones. Tape recording, August 9.

Maryland Sustainable Growth and Preservation Act. 2012. S.B. 236, Maryland Regular Session, Md. Laws 149 (enacted).

Mazmanian, Daniel, and David Morell. 1992. *Beyond Superfailure: America's Toxics Policy for the 1990s*. Boulder, CO: Westview Press.

McCabe, Robert. 2010. Norfolk loses cruise line as Baltimore steams ahead. *Virginian-Pilot* (Norfolk, VA), June 13.

McCarthy, Gerald. 2012. Interview with Shana Jones. Tape recording, October 19.

McGinnis, Michael V. 1999. Making the watershed connection. *Policy Studies Journal* 27:497–501.

McGuire, Michael. 2006. Collaborative public management: Assessing what we know and how we know it. *Public Administrative Review* 66 (Special Issue, December): 33–43.

McKinney, Matthew, and Patrick Field. 2008. Evaluating community-based collaboration on federal land and resources. *Society and Natural Resources* 21:419–29.

McKinney, Matthew, and William Harmon. 2004. *The Western Confluence: A Guide to Governing Natural Resources*. Washington, DC: Island Press.

McLeod, John. 2012. Telephone interview with William Gibson, December 15.

McNamara, Madeleine. 2012. Starting to untangle the web of cooperation, coordination, and collaboration: A framework for public managers. *International Journal of Public Administration* 35:389–401.

_____. 2010. Collaborative environmental management within a traditionalistic political culture: An unconventional approach to "wicked" problems. In *Speaking Green with a Southern Accent: Environmental Management and Innovation in the South*, ed. Gerald A. Emison and John C. Morris, 129–48. Lanham, MD: Lexington Books.

McNamara, Madeline W., and John C. Morris. 2012. More than a one-trick pony: Exploring the contours of a multi-sector convener. *Journal of Non-Profit Management* 15: 84-103.

McNamara, Madeleine, William M. Leavitt, and John C. Morris. 2010. Multiple-sector partnerships and the engagement of citizens in social marketing campaigns: The case of Lynnhaven River Now. *Virginia Social Science Journal* 45:1–20.

Meyer, Stephen, and David Konsiky. 2007. Adopting local environmental institutions: Environmental need and economic constraints. *Political Research Quarterly* 60:3–16.

Moore, Elizabeth, and Tomas Koontz. 2003. A typology of collaborative watershed groups: Citizen-based, agency-based, and mixed partnerships. *Society and Natural Resources* 16:451–60.

Morris, John C. 1997. The distributional impacts of privatization in national water quality policy. *Journal of Politics* 59:56–72.

_____. 1996. Institutional arrangements in an age of New Federalism: Public and private management in the State Revolving Fund program. *Public Works Management & Policy* 1:145–57.

Morton, Lois. 2008. The role of civic structure in achieving performance-based watershed management. *Society and Natural Resources* 21:751–66.

Moses, Ray. 2012. Interview with William Gibson. Tape recording, March 22.

_____. 1996. Elizabeth River is far from cleansed. *Virginian-Pilot* (Norfolk, VA), May 30.

Mudflats. 2012a. *VA Port Authority Earns Hall of Fame for Reducing Pollution*, p. 4.

_____. 2012b. *The Best Lawn Ever*, p. 5.

Nansemond River Preservation Alliance (NRPA). 2012a. River history. nansemondriversavers.org/river-news.

_____. 2012b. Nansemond River Preservation Alliance mission statement. www.nansemondriversavers.org/our-mission.

_____. 2012c. Nansemond River Preservation Alliance Officers. nansemondriversavers.org/contact-us.

_____. 2011. 2010 State of the Nansemond report. Suffolk, VA: Nansemond River Preservation Alliance Water Quality Committee. www.nansemondriversavers.com/images/stories/reports/2010StateNansemondRiverReportJune142011Final.pdf.

_____. 2010. *Bylaws of the Nansemond River Preservation Alliance*. Suffolk, VA: Author.

National Oceanic and Atmospheric Administration (NOAA). 2012. About us. Chesapeake Bay Office, Norfolk.

Nelson, Lisa. 1999. Environment in the community, community in the environment. *Administrative Theory & Praxis* 21:354–61.

Nelson, Lisa, and Lois Weschler. 2001. The watershed as a focus for public administration: Conventional and bioregional approaches. *Administrative Theory & Praxis* 23:10–24.

Newport News Department of Development. 2011. *The Seafood Industrial Park*. www.nngov.com/development/resources/sip.html.

Nickelsberg, Mike. 2012. Interview with William Gibson. Tape recording, April 14.

Norfolk Department of Development. 2011. *Military and Defense Industry*. www.norfolkdevelopment.com/index.php?option=com_content&view=article&id=139&Itemid=55.

Norfolk Naval Shipyard. 2012. *NNSY History*. navsea.navy.mil/shipyards/norfolk/History/Home.aspx.

O'Brien, Dennis. 2001. Proposed Portsmouth, VA marine terminal could disrupt wetlands. *Virginian-Pilot* (Norfolk, VA), June 2.

Oliver, James. 2012. Interview with William Gibson. Tape recording, September 12.

Ostrom, Elinor. 1990. *Governing the Commons: The Evolution of Institutions for Collective Action*. New York: Cambridge University Press.

———. 1986. A method of institutional analysis and an application to multiorganizational arrangements. In *The Public Sector: Challenge for Coordination and Learning*, ed. Franz-Xaver Kaufmann, 501–23. Berlin: Walter deGruyter.

Parisi, Domenico, Michael Taquino, Steven M. Grice, and Duane A. Gill. 2004. Civic responsibility and the environment: Linking local conditions to community environmental activeness. *Society and Natural Resouces* 17:97–112.

Parker, K., Richard Margerum, D. Dedrick, and J. Dedrick. 2010. Sustaining watershed collaboratives: The issue of coordinator-board relationship. *Society and Natural Resources* 23:469–84.

Peace, Whit. 2011. Why I do river cleanups. In *The Lynnhaven: Restoring a Legend*, 23. Virginia Beach, VA: Lynnhaven River NOW.

_______. 2012. Interview with William M. Leavitt. Tape recording, March 29.

Picardi, Mary. 2012. Interview with William M. Leavitt. Tape recording, April 4.

Policy Consensus Initiative. 2012. What is collaboration governance? *Policy Consensus Initiative*. July 22. www.policyconsensus.org/publicsolutions/ps_2.html.

Poncelet, Eric. 2004. *Partnering for the Environment: Multistakeholder Collaboration in a Changing World*. Lanham, MD: Rowman & Littlefield.

Port of Virginia. 2012. *Green Operator*. www.portofvirginia.com/environment/green-operator.aspx.

———. 2011. *Economic development*. Norfolk, VA: Author.

Priest, Josh. 2012. Interview with William Gibson. Tape recording, March 27.

Priest, Walter. 2012a. Interview with William Gibson. Tape recording, April 12.

———. 2012b. Interview with William Gibson. Tape recording, July 26.

Purdy, Jill. 2012. A framework for assessing power in collaborative governance processes. *Public Administration Review* 72:409–17.

Putnam, Robert D. 2000. *Bowling Alone: The Collapse and Revival of American Community*. New York: Simon & Schuster.

Randalph, John, and Michael Bauer. 1999. Improving environmental decision-making through collaborative methods. *Policy Studies Review* 16:168–91.

Reisch, Timothy. 2012. Interview with William Gibson. Tape recording, July 16.

Rieger, Joe. 2012. Interview with William Gibson. Tape recording, June 15.

Rittel, Horst W. J., and Melvin M. Webber. 1973. Dilemmas in a general theory of planning. *Policy Sciences* 4:155–69.

Ritter, Ron. 2012. Interview with William Gibson. Tape recording, June 18.

Robinson, Catherine J., Richard D. Margerum, Tomas M. Koontz, Cassandra Moseley, and Sue Lurie. 2011. Policy-level collaboratives for environmental management at the regional scale: Lessons and challenges from Australia and the United States. *Society and Natural Resources* 24:849–59.

Rosenbaum, Walter A. 2002. *Environmental Politics and Policy*, 5th ed. Washington, DC: CQ Press.

Rule, Joe. 2012. Interview with William Gibson. Tape Recording, December 3.

Ryan, Louis F., and Andrew S. Fine. 2011. Don't impede bay cleanup plan. *Virginian-Pilot* (Norfolk, VA), March 31.

Sabatier, Paul, Christopher Weible, and Jared Ficker. 2005. Eras of water management in the United States: Implication for collaborative watershed approaches. In *Swimming Upstream: Collaborative Approaches to Watershed Management*, by Paul Sabatier, W. Focht, M. Lubell, Z. Trachtenberg, A. Vedlitz, and M. Matlock, 23–51. Cambridge, MA: MIT Press.

Salamon, Lester M. 2002. *The Tools of Government: A Guide to the New Governance*. New York: Oxford University Press.

Salzman, James, and Barton H. Thompson, Jr. 2012. *Environmental Law and Policy*, 3rd ed. New York: Foundation Press.

Schuett, Michael A., Steve W. Selin, and Deborah S. Carr. 2001. Making it work: Keys to successful collaboration in natural resource management. *Environmental Management* 27:587–93.

Serpe, Richard. 2012. Interview with William M. Leavitt. Tape recording, April 11.

Shaw, Mary. 2003. Successful collaboration between the nonprofit and public sector. *Nonprofit Management & Leadership* 14:107–20.

Sheler, Jeff. 2012. Fourteen million gallons and counting: Sewage spills into Suffolk creek. *Virginian-Pilot* (Norfolk, VA), November 6.

Simmel, George. 1950. *The Sociology of George Simmel*. New York: Free Press.

Sirianni, Carmen. 2009. The civic mission of a federal agency in the age of networked governance: U.S. Environmental Protection Agency. *American Behavior Scientist* 52:933–52.

Smith, Courtland, and Jennifer Gilden. 2002. Assets to move watershed councils from assessment to action. *Journal of the American Water Resources Association* 38:653–62.

Smith, Karla. 2012a. Interview with John C. Morris. Tape recording, June 20.

_____. 2012b. Interview with John C. Morris. Tape recording, April 20.

_____. 2012c. Interview with John C. Morris. Tape recording, March 5.

Smith, Steven R., and Helen Ingram. 2002. Policy tools and democracy. In *The Tools of Government*, by Lester Salamon, 565–84. New York: Oxford University Press.

Sorabella, Laurie. 2012. Interview with William M. Leavitt. Tape recording, May 24.

_____. 2011. Goal: 100 acres of oyster habitat. In *The Lynnhaven: Restoring a Legend*, 14. Virginia Beach, VA: Lynnhaven River NOW.

Spore, James. 2012. Interview with William M. Leavitt. Tape recording, April 9.

Stanton, Robert. 2012. Interview with William M. Leavitt. Tape recording, March 14.

Stedfast, Phil. 2012. Interview with William Gibson. Tape recording, April 15.

Steelman, Toddi A., and Jorge Rivera. 2006. Voluntary environmental programs in the United States: Whose interest are served? *Organization & Environment* 19:505–26.

Stiles, Skip. 2012. Interview with Shana Jones. Tape recording, February 17.

Swift, Earl. 1993. From lean beginnings, movement to save river is growing. *Virginian-Pilot* (Norfolk, VA), October 3.

Taraski, Elizabeth. 2012a. Interview with John C. Morris. Tape recording, February 17.

_____. 2012b. Interview with John C. Morris. Tape recording, April 25.

_____. 2012c. Interview with John C. Morris. Tape recording, June 20.

Taraski, Elizabeth, Karla Smith, and John Wass. 2012. Interview with John C. Morris. Tape recording, January 27.

Thomas, Craig. 2008. Evaluating the performance of collaborative environmental governance. *Consortium on Collaborative Governance Mini-conference*. Santa Monica, CA: University of Washington.

Thomas, Joseph. 2012. Interview with William Gibson. Tape recording, Mar. 23.

Thompson, James D. 1967. *Organizations in Action: Social Science Bases of Administrative Theory*. New Brunswick, NJ: Transaction Publishers.

Thomson, Ann Marie. 2001. *Collaboration: Meaning and Measurement*. PhD diss., Indiana University–Bloomington.

Thomson, Ann Marie, and James L. Perry. 2006. Collaboration process: Inside the black box. *Public Administration Review* 66 (Special Issue, December): 20–32.

de Tocqueville, Alexis. 1997 (1835; *tr.* 1899). *Democracy in America*. Electronic edition deposited and marked-up by ASGRP, the American Studies Programs at the University of Virginia. xroads.virginia.edu/~HYPER/DETOC/ch2_08.htm.

Trachtenberg, Zev, and Will Focht. 2005. Legitimacy and watershed collaborations. In *Swimming Upstream: Collaborative Approaches to Watershed Management*, by Paul Sabatier, Will Focht, Mark Lubell, Zev Trachtenberg, and Marty Matlock, 53–82. Cambrige, MA: MIT Press.

Travis, Rick, John C. Morris, and Elizabeth D. Morris. 2004. State implementation of federal environmental policy: Explaining leveraging in the Clean Water State Revolving Fund. *Policy Studies Journal* 32:461–80.

Tribo, Jenny, and Whitney Katchmark. 2012. Interview with Shana Jones. Tape recording, February 28.

Unger, Michael. 2012. Interview with William Gibson. Tape recording, June 27.

University of Virginia. 2005. A Brief History of Money Point. www.virginia.edu/ien/moneypoint/docs/Money_Point_history.pdf.

University of Virginia Institute for Environmental Negotiation. 2005. *Collaborating around Contamination: Revitalizing the Money Point Corridor*. www.virginia.edu/ien/moneypoint/mpointfiles/Money%20Point%20background%20paper.pdf.

US Census Bureau. 2012. 2010 Census interactive population search. 2010.census.gov/2010census/popmap/ipmtext.php?fl=51.

_____. 2010. *Population and Housing Occupancy Status: 2010—United States—Metropolitan Statistical Area*.

_____. 1982. 1980 Census of population: Volume 1, Characteristics of the Population: Chapter B, General Population Characteristics: Part 48, Virginia. Washington, DC: Department of Commerce. www2.census.gov/prod2/decennial/documents/1980/1980censusofpopu80148uns_bw.pdf.

US Environmental Protection Agency (USEPA). 2012a. *National Priorities List*. www.epa.gov/superfund/sites/npl.

_____. 2012b. About the EPA: EPA History (1970–1988). www.epa.gov/aboutepa/history/topics/epa/15b.html.

_____. 2012c. *Region 3: Elizabeth River Team*. www.epa.gov/region03/elizabethriver/elizabethriver_team.html.

_____. 2012d. *Total Maximum Daily Loads (TMDLs) at Work*. water.epa.gov/lawsregs/lawsguidance/cwa/tmdl/upload/2009_12_11_tmdl_tmdlsatwork_lynnhaven_river_sound_byte.pdf.

_____. 2012e. Water: Adopt your watershed. water.epa.gov/action/adopt/index.cfm.

_____. 2010a. *Final Chesapeake Bay Total Maximum Daily Load for Nitrogen, Phosphorus, and Sediment*.

_____. 2010b. *Polluted Run Off (Nonpoint Source)*. www.epa.gov/owow_keep/NPS/index.html.

_____. 2008. Surf your watershed. yosemite.epa.gov/water/surfnote.nsf/FTsearchForm?readform&Limit=200&Query=02080108.

_____. 1995. Toward sustainable ecosystems and communities: Community-based environmental protection. nepis.epa.gov/Exe/ZyNET.exe/400001F0.txt?ZyActionD=ZyDocument&Client=EPAamp;Index=1995%20Thru%201999&Docs=&Query=FNAME%3D400001F0.

TXT%20or%20%28%20people%20or%20places%20and%20partnerships%29&Time=amp;EndTime=&SearchMethod=1&TocRestrict=n&Toc=&TocEntry=&.

US Fish and Wildlife Service. 2012. Getting rid of the "goo": Restoring and monitoring Elizabeth River. www.fws.gov/northeast/ecologicalservices/fap/vafapelizabethriver.html.

US Internal Revenue Service. 2010. Form 990.

Vig, Norman J., and Michael E. Kraft. 2000. Environmental policy from the 1970s to 2000: An overview. In *Environmental Policy*, 4th ed., ed. Norman J. Vig and Michael E. Kraft, 1–31. Washington, DC: CQ Press.

Virginia Beach Economic Development. 2012. *Economic Sectors*. www.yesvirginiabeach.com/business-data/pages/economic-sectors.aspx.

Virginia Conservation Network. 2012. About us. vcnva.org/anx/.

Virginia Department of Environmental Quality (DEQ). 2012a. Superfund. www.deq.virginia.gov/Programs/LandProtectionRevitalization/RemediationProgram/Superfund.aspx.

_____. 2012b. Chesapeake Bay assessment. www.deq.state.va.us/Portals/0/DEQ/Water/WaterQualityAssessments/IntegratedReport/2012/ir12_Ch4.4_CBP_Assessment.pdf.

_____. 2004. Lynnhaven Bay, Broad Bay, and Linkhorn Bay Watersheds: Total Maximum Daily Load (TMDL) Report for Shellfish Areas Listed Due to Bacteria Contamination. Richmond: VA.

Virginia Port Authority. 2012. About. www.portofvirginia.com/About.aspx.

Virginian-Pilot. 2011. Time to tend to the Nansemond. June 26.

_____. 2003. Countdown begins on the Lynnhaven River. April 19.

Wagner, Cheryl L., and Maria Fernandez-Gimenez. 2008. Does community-based collaborative resource management increase social capital? *Society and Natural Resources* 21:324–44.

Warner, William W. 1976. *Beautiful Swimmers: Watermen, Crabs, and the Chesapeake Bay*. New York: Little, Brown.

Washington State Department of Natural Resources. 2008. *Brief Science of Creosote*. www.dnr.wa.gov/Publications/aqr cleanup creosote brief.pdf.

Wass, John. 2012a. Interview with John C. Morris. Tape recording, March 5.

_____. 2012b. Interview with John C. Morris. Tape recording, July 3.

Wayland, Robert. 2012. Interview with John C. Morris. Tape recording, May 16.

Weber, Edward. 2009. Explaining institutional change in tough cases of collaboration: "Ideas" in the Blackfoot Watershed. *Public Administration Review* 69:314–27.

_____. 2000. A new vanguard for the environment: Grass-roots ecosystem management as a new environmental movement. *Society and Natural Resources* 13:237–59.

Wiersema, Annecoos. 2008. A train without tracks: Rethinking the place of law and goals in environmental and natural resources law. *Environmental Law* 38:1239–81.

Wilson, James Q., and Edward Banfield. 1971. Political ethos revisited. *American Political Science Review* 65:1048–62.

_____. 1964. Public-regardingness as a value premise in voting behavior. *American Political Science Review* 58:876–87.

_____. 1963. *City Politics*. Cambridge, MA: Harvard University Press.

Wondolleck, Julia, and S. Yaffee. 2000. *Making Collaboration Work: Lessons from Innovation in Natural Resource Management*. Colvelo, CA: Island Press.

Wood, Donna J., and Barbara Gray. 1991. Toward a comprehensive theory of collaboration. *Journal of Applied Behavioral Science* 27:139–62.

Wood, Heather. 2012. Interview with William Gibson. Tape recording, July 17.

Worth, Nicholas. 2012a. Interview with John C. Morris. Tape recording, March 23.

_____. 2012b. Interview with John C. Morris. Tape recording, June 29.

Wyatt, Dottie. 2012. Interview with William Gibson. Tape recording, July 17.

Yarsinske, Amy Waters. 2007. *The Elizabeth River*. Charleston, SC: The History Press.

Yin, Chava-Frankfort, and David Nachmias. 1996. *Research Methods in the Social Sciences*, 5th ed. New York: St. Martin's Press.

Yin, Robert K. 1994. *Case Study Research: Design and Methods*, 2nd ed. Thousand Oaks, CA: Sage.

Index

Advisory Board, 95
Alperin, Brian, 203
APM Terminals, 89, 107, 125, 127

BAE Systems, 120, 121, 122, 130, 138, 142n4
Barnett, Cheryl, 102, 136
Barnum, Steven, 62, 192
Barrow, Mary Reed, 145, 164, 173
Barry, Richard "Dick," 71, 187, 188, 189, 190, 192, 193, 194, 195, 198, 201, 202, 207, 209, 216n9; "Dick Barry Rule," 198
Bennett's Creek Park, 191, 196
Benthic Community, 125
Bernick, Clay, 60, 154, 158, 176
Best management practices (BMPs), 39, 79; Adoption of, 39, 79, 229
BIMBY (Because It's My Back Yard), 13, 16, 17n7, 78, 82, 122, 226, 229, 231
Birdsong, Dr. Ray, 117
Birdsong Wetlands, 117, 136, 139
Boatwright, Pamela, 105, 111, 119, 120, 121
Bradley, Bruce, 106, 107, 136
Bray, Robert, 90, 105
Browning, Wilson, 189, 200
Business Partners for a Cleaner River, 119

Calcium Hydroxide, 92, 138
Captain John Smith, 89
Carson, Rachel: *Silent Spring*, 24
Chalmers, Hap, 143, 152, 153, 174
Chesapeake Bay, xv, xvi, xvii, xx, xxiv, 1, 2, 5, 9, 10, 11–12, 14–15, 16, 17n5, 33, 54, 55, 56, 57, 60, 61, 64, 65, 66, 69, 72, 75, 77, 78, 79, 85, 87, 89, 91, 93, 132, 133, 144, 156, 184, 186, 205, 207–208, 217, 219
Chesapeake Bay Foundation, 55, 60, 69, 87, 148, 189, 190, 222
Chesapeake Bay Preservation Act, 93, 202
Chesapeake Bay Program, 33, 58, 60, 64, 66
Choose Clean Water Coalition (CCWC), 69, 73, 222
Chuckatuck Creek, 186–187, 189, 192, 197, 201, 208, 211
Citizens for a Cleaner River, 119
City of Chesapeake, 66, 126, 127
City of Norfolk, 62, 70, 82, 107, 130, 132
Clean the Bay Day, 70, 71, 93, 94
Clean Water State Revolving Loan Fund, 25
Coastal Zone Management Act of 1972, 25
Coccaro, Joseph, 96
Collaboration: Definition of, 5, 6; Linkages with national and regional groups, 53, 64, 75, 78
Colonna's Shipyard, 122
Comparative Risk Analysis, 96; Comparative Risk Analysis Branch, 96, 105
Comparative Risk Assessment, 97, 99, 105, 109, 110, 115
Comprehensive Environmental Response, Compensation, and Liability Act (CERCLA), 9
Constant's Wharf, 201
Context: Definition, 22; Summary, 219, 223, 227
Convener: Definition of, 31
Copeland, Howard, 62, 75
Corbin, Jeff, 64, 68, 78, 79, 83
Craddock, 91, 135
Creosote, 85, 125
Crisman, Phoebe, 129, 130

Crocker, Bob, 192, 201, 211
Cross, Katherine, 94

Daniel, Frank, 65, 101, 108, 114, 125
Dave Iwans and Associates, 149, 159
Davis-Martin, James, 66, 72, 73, 79
Dean, Robert, 70, 74, 93
Dissolved Oxygen Meter: Measurement, 201; Purchase, 195, 201; Purpose, 195
Dollar Tree Corporation, 152, 159, 169, 177
Dunbar, Robin, 128, 129, 130

Eclipse, 183, 186, 216n5
Elazar, Daniel, xxi, xxiii, 8, 27, 28, 51n5, 86, 146, 186. *See also* Political Culture
Elizabeth River, xiii, 10, 17n5, 50, 53, 54, 57, 60–61, 63, 65, 67, 70, 74, 82, 83, 85, 86, 86–87, 87, 88, 89, 90, 92, 94, 95–96, 96, 98, 99, 100, 101, 103, 104, 107, 108, 110, 115, 115–116, 124, 130, 131, 135, 142n1, 142n3, 184, 186, 203, 219, 220, 222, 226, 227, 228
Elizabeth River Project, xiii, xix, 50, 55, 60, 86, 92, 96, 101, 108, 112, 121, 140, 148, 187, 203, 219; Conveners, 104, 106, 140; Elizabeth River Restoration Trust Operating Agreement, 107, 125; Elizabeth River Visions, 96, 110; Executive Committee, 111; First Watershed Action Plan, 99; Fundraising Committee, 112, 115; Governance Structure, 110; Governing Board/ Project Team, 95–96, 105, 110; Lafayette River TMDL, 82; Mission Statement, 96, 104, 118; Public Relations Committee, 111–112; River Restoration Advisory Committee, 119–120; River Stars, 119–120, 121, 123, 131, 142, 222; River Star Businesses, 105, 107, 118, 119, 120, 121, 123, 124, 128, 130, 135; River Star Hall of Fame, 98, 122; River Star Homes, 104, 128, 131–132, 222, 226; River Star Schools, 101, 128, 129; River Stars – P2 Workbook, 120; Sediment Remediation Partnership Committee, 115; Technical Programs Committee, 113, 114; The Learning Barge, 128, 129, 130; Watershed Action Plan, 86, 95, 99, 100, 106, 109, 110, 111, 113, 116, 117, 119, 124, 125, 133, 140; Watershed Action Team, 99, 136
Endangered Species Act, 24
Eppinger and Russell, 124
Ernst, Howard, xix, 10, 77
Eure, Jack, 188, 198, 202, 203–204, 212
Everett, Christy, 66, 68, 71, 72, 81, 162, 189
Ewing, Michael, 120, 121, 122, 138

Fine, Andrew, 71, 147, 148, 149, 157, 159, 178, 188
Forget, Karen, xiii, 72, 73, 80, 81, 143, 145, 151–152, 153, 157, 159, 160, 161, 164, 168, 169, 170, 171, 175, 177
Frankenfield, Barry, 149–150, 157, 168

Gates, Janice, 206
General Revision Act of 1891, 23
Godwin Bridge, 199
Gosport Naval Yard, 89
Great Bridge, 87
Great Dismal Swamp, 200

Hampton Roads, xx, xxiii, 3, 9, 10, 53–55, 56, 57, 59, 60, 61, 62, 63, 64–65, 65, 66–68, 69, 69–72, 79, 80, 81, 82, 83, 87, 89, 89–90, 93, 96, 104, 120, 139, 142n3, 146, 153, 162, 178, 180, 184, 186, 188, 219, 232n1; Location of, xx, 3, 10, 54; Port of, 89. *See also* Tidewater Virginia
Hampton Roads Planning District Commission (HRPDC), 9, 59, 60, 66, 67–68, 92, 200
Hampton Roads Sanitation District (HRSD), 104, 119, 123, 131, 132, 200
Harper, Scott, 70, 123
Herbert, Steve, 154, 156
Herbert, Todd, 72
Hill, Noah, 72
Hinshelwood, Geoff, 192, 202
Host, Mike, 62, 105, 111, 133, 136
Huggett, Dr. Robert, 96, 105, 110

Hume, Bob, 114, 125, 126
Hurricane Isabel, 211

Inside Business: The Hampton Roads Business Journal, 120
Institute for Environmental Negotiation, 93
Internal Revenue Service, 158, 189; 501(C)(3) status, 49; Form 990, 158, 195–196, 216n7
Izzak Walton League, 194

Jackson, Marjorie Mayfield, xiii, 61, 65, 74, 84n1, 92–93, 96, 97, 104, 105, 105–106, 108, 109, 110, 111, 120, 124, 127, 130, 132, 139, 142, 142n2, 187
James River, 53, 55, 57, 58, 74, 87, 89, 184, 185, 186, 197

Katchmark,Whitney, 67, 68, 69
Kempe's Landing, 88
Kempsville, 88
Kensler, Michael, 70, 71, 74, 93
Kinder Morgan/Elizabeth River Terminals, 120, 122, 126, 129
King James, 89
Koubsky, Dave, 101, 134, 135
Kuralt, Charles, 99, 100

Lafayette River, 50, 62, 70, 82, 87, 88, 89, 104, 107, 117, 130, 132, 135, 220; Lafayette River Restoration Project, 130; Lafayette Riverfest, 131, 216n11
Lambert's Point Coal Terminals, 89
Land and Water Conservation Act, 24
Landmark Communications, 187, 200
Larchmont Library, 117
League of Women Voters, 93
Legacy Pollutants, 91, 142n1
Lester, Harry, 71, 74, 147, 157, 178, 188
Linkenhoker, Cindy, 192
Living River Restoration Trust, 107, 125
London Company, 89
Lott, Nathan, 72, 73
Lynnhaven Oyster Company, 143, 152, 153
Lynnhaven River, 143, 144, 145, 146–147, 147, 148, 150, 152, 153, 154, 155, 156, 157, 158, 162, 163, 164, 165, 166, 167, 169, 170, 171, 172, 173, 174, 175, 177, 178, 179–180, 184, 188, 211, 214, 219, 220
Lynnhaven River NOW, xiii, xix, 55, 56, 60, 143, 144, 149, 154, 158, 161, 162, 164, 179, 181, 184; Annual Budget, 151; Annual Oyster Roast, 143, 152, 177, 180; Bylaws, 160, 161; Conveners, 157, 179; Lynnhaven River Now Committee Structure, 150–151, 157, 160, 161, 180, 181; Clean Boating and Clean Marina Committee, 162; Education Committee, 163; Executive Program Committee, 151, 160, 161; Landscape Practices Committee, 161, 163; Oyster Restoration Committee, 151, 160, 161, 163; Public Policy Committee, 161, 164, 173; Public Relations and Marketing Committee, 159, 161, 164, 173; Stewardship and Access Committee, 161, 165; Wetlands Committee, 166, 172; Lynnhaven River 2007, 146, 148, 153, 174, 179; Oyster Restoration, 60, 72; Pearl Homes, 159, 168, 170, 171, 222, 226; Pearl Schools, 159, 163, 170, 171; Stakeholders, 150, 152, 153, 154, 155–157; State of the River Report, 164, 170; Wetlands in the Classroom Program, 169

Martin, Howard, 188, 189
McCarthy, Jerry, 74, 94
McLeod, John, 132
Money Point, 85–86, 87, 101, 124, 126, 127–128, 129, 132, 134, 135, 137, 141, 222; Money Point Revitalization Task Force, 126
Moore, Chris, 200
Moses, Admiral Ray, 62, 106, 111, 118, 121, 125
Multiple-case design, 47, 48
Mummichog, 96, 125, 135

Nansemond River, 53, 54, 55, 57; Lake Mead, 185, 211; Location and

characteristics, 57, 200; Shingle Creek, 215; Watermen, 193, 212–213, 215, 216n13
Nansemond River Power Squadron, 194
Nansemond River Preservation Alliance: Conveners, 194–195; Education Committee, 198, 205, 210; Governance Committee, 198, 204; Government Relations Committee, 204; Mission statement, 189, 189–xxv, 207, 214; Oyster Restoration Committee, 199; River and Creek Fest 2012, 191, 194, 196, 204, 206, 210; John Smith Water Trail map, 191, 197, 207–208; Shoreline Committee, 198, 202, 203, 210; Structure, 197, 199–204; Water Quality Committee, 190, 200–202, 205, 210
Nansemond Suffolk Academy, 204–205
Nansemond Tribe, 208
National Environmental Policy Act, 25
National Oceanographic and Atmospheric Administration (NOAA), 53, 59, 60, 61–62, 65, 66, 70, 82, 106, 114, 127, 130, 183, 192
National Park Service, 204, 207
National Priority List, 9
Nauticus, 99, 110, 117
Neikirk, Chris, 107, 115
Nickelsberg, Mike, 105, 114, 128
NIMBY (Not In My Back Yard), 12–13, 226
Norfolk International Terminals, 89
Norfolk Naval Shipyard, 54, 62, 89, 91, 105, 123, 133, 136
Norfolk Naval Station, 89, 92

Oak Ridge Center for Risk Analysis, 98
Oyster Restoration, 60, 72, 150, 151, 156, 158, 160, 161, 163–164, 169

Paradise Creek Nature Park, 87, 91, 101, 106, 107, 115, 120, 128, 132, 133, 223
Paradise Found, 133
Pleasure House Point, 150
Political culture. *See also* Elazar, Daniel: Chesapeake, 87; Definition of, xxi, 27; Norfolk, 86–87; Portsmouth, 86–87; Role in collaborations, 55, 219–220, 223, 227; Suffolk, 186; Types, xxiii, 27–29, 86; Virginia Beach, 146
Pollutants: Agricultural Runoff, 55, 102; Bacteria, 55, 104, 113, 130, 132, 146, 170, 174, 185, 200–201, 215; Nutrient (Nitrogen and Phosphorus), 55, 130, 170; Polycyclic Aromatic Hydrocarbons (PAHs), 57, 85; Stormwater, 55, 102, 113, 130, 210
Potomac River, 15
Priest, Josh, 117, 118, 139
Priest, Walter, 62, 66, 114, 121, 127, 183, 202, 206
Princess Anne, 88
Princess Elizabeth, 89, 128
Progressive Era, 23
Putnam, Robert: Bowling Alone, 229

Reclamation Act of 1902, 23
Refractometer, 195, 201
Reisch, Tim, 138, 142
Rieger, Joe, 101, 105, 108, 114, 134, 137
Ritter, Ron, 121, 138
River and Harbors Act of 1925, 23
River Talks, 206, 210
Rule, Joe, 114, 122
Runnymede Corporation, 147, 149, 159
Ruritans, 186; Ruritan Hall, 192

Scott's Creek, 92, 139
Secchi disk, 196, 202
Skipjack, 211
Smith, Karla, 187, 188–189, 195, 198, 199
Snowball sampling, 48
Social capital, xx, xxi, xxiii, 3, 4, 20, 45, 97, 99, 100, 110, 115, 135, 138; Comparison across cases, 221, 222–223, 224–225, 226, 229; Components, 6, 8, 21, 22, 26, 29, 39–40, 41, 42, 43, 46; Definition, 7, 42, 135; Examples of, 31, 36, 41, 134, 136, 138; Lessons of, 140, 141, 218;

Motivation for BIMBY, 13, 15, 17n7, 225, 226
Sorabella, Laurie, 150–151, 158, 163, 169, 172, 178, 180
Southeastern Association for Virginia's Environment, 93
Southeastern Public Service Authority, 134
Spat, 164, 172, 182n3, 196
Spore, James, 146, 147, 148, 149, 157, 175
Stanton, Robert, 143, 147, 148, 157, 175, 177
Stedfast, Phil, 85, 120–121, 122, 129
Stiles, Skip, 70, 73, 82, 83
Suffolk (VA), City of: Departments, 194, 197, 203; Department of Economic Development and Tourism, 194, 207; Description, 184–185; Office of Media and Community Relations, 186, 197; Planning Commission, 215; Wetlands Board, 192, 197, 202, 203
Suffolk Master Gardeners, 192, 210
Suffolk River Heritage, 186–187, 195, 204, 207–208, 209
Superfund (see CERCLA), 9, 17n4, 17n5, 62, 65
Susquehanna River, 13–14, 58
Systems Model, 20, 40, 47

Taraski, Elizabeth, 74, 190, 192–193, 193–194, 195, 198, 199, 209
Tennessee Valley Authority (TVA), 23
Thomas, Joe, 60, 62, 106, 112–113, 114, 125, 128
Tidewater Community College, 104, 114
Tidewater Virginia, 10. *See also* Hampton Roads
Total Maximum Daily Load (TMDL), xxii, 9, 16, 205, 230; Bay TMDL, 63, 66, 67, 68, 70, 71–72, 73, 77, 78–79, 83, 84n2, 84n4, 219, 229; Lafayette River TMDL, 82; Lynnhaven River TMDL, 80, 80–81, 83; Process, 63, 66, 67, 69, 80, 82
Tribo, Jennifer, 60, 67, 68, 69, 82
Tributyltin, 123

Unger, Michael, 92, 114, 122, 124, 127, 135
University of Virginia, 93, 124; Institute for Environmental Negotiation, 93; School of Architecture, 129
U.S. Army Corps of Engineers, 53, 59, 60–61, 61, 62, 67, 82, 101, 106, 107, 114, 125, 126, 142n5, 148, 151, 153, 156, 158, 160, 164, 170, 172, 221
U.S. Coast Guard Support Center (Portsmouth, VA), xx, 54, 98
U.S. Environmental Protection Agency, 24, 37, 59, 61, 63, 121, 122; Chesapeake Bay Agreement, 76; Clean Water Act, 84n2; Collaborative arrangements, 64, 79, 83, 101, 103, 126, 127, 137; Delegation, 64; Formation of, 24, 25, 25–26; Office of Policy, Planning and Evaluation, 96, 105, 110; Science Advisory Board, 96, 105; Superfund List, 90, 142n5; Technical Support, 63; TMDL Process, 63, 66, 68, 77, 80, 81, 82; Toxic Remediation Team, 57, 63, 65, 90
U.S. Forest Service, 23; Example of VEP, 37
U.S. Navy, xx, 54, 91, 94, 102, 118, 133, 138
USS *Merrimack*, 91

Virginia Beach, City of, 54, 55, 56, 57, 60, 61, 70, 87–88, 92, 144, 156, 219–220, 221, 229; Clean Water Task Force, 152, 156, 161, 169; Consolidation, 145; Green Ribbon Committee, 156, 161, 168, 169; Oyster Heritage Fund, 152, 159; Partnership with, 148, 156, 168, 174; SPCA, 93; Storm Water Utility Fund,; TMDL, 80–81
Virginia Coastal Zone Management Program, xxiii
Virginia Conservation Network, 60, 69, 72–74
Virginia Department of Conservation and Recreation, 59, 60, 65–66, 72, 73, 83, 134

Virginia Department of Environmental Quality (DEQ), 59, 63, 64–65, 67, 80, 91, 101, 107, 108, 114, 119, 125, 126, 127, 132, 142n5, 200, 200–201, 205
Virginia Department of Health, 56, 132, 173, 200
Virginia Environmental Endowment, 60, 69, 74, 94, 97, 106, 204
Virginia Institute of Marine Science, 60, 61, 82, 96, 101, 103, 105, 110, 113, 114, 122, 124, 127, 150, 156, 164, 180, 221
Virginia Land Conservation Foundation, 134
Virginia Port Authority, 55, 90, 105, 107, 120, 132, 134
Virginia Ship Repair Association, 121
Virginian-Pilot, 50, 68, 70, 92, 96, 106, 111, 123, 134, 164, 187
Voluntary Environmental Programs (VEPs): Definition of, 36
Volvo-Penta Test Center, 191, 192, 201, 211, 231

Wass, John, 188–189, 198, 199
Water Pollution Control Act of 1972 (Clean Water Act), 24, 25, 53, 63, 64, 66, 76, 76–77, 80, 84n2
Water Quality Act of 1965, 24, 51n4
Water Quality Act of 1987, 25
Wayland, Robert, 64, 71, 83
Wetlands Loss, 55, 57
Wicked problems, 10, 217, 230, 231; Definition, 8–9
Wilderness Act, 24
Witchduck Road, 88
Wood, Heather, 120
Wood, John, 89, 90
Worth, Nicholas, 65, 183, 190, 195, 200–202, 209
Wyatt, Dottie, 135

About the Authors

Dr. John Charles Morris is a professor of public administration and PhD graduate program director in the Department of Urban Studies and Public Administration at Old Dominion University in Norfolk, Virginia. He has studied the Clean Water State Revolving Fund program since its inception in 1987, and has published widely in public administration and public policy. Dr. Morris has a significant number of publications in environmental policy and politics. He is the coeditor of *Speaking Green with a Southern Accent: Environmental Management and Innovation in the South* (Lexington 2010), and *True Green: Executive Effectiveness in the US Environmental Protection Agency* (Lexington 2012). Professor Morris is also the coeditor of *Building the Local Economy: Cases in Economic Development*, published by the Carl Vinson Institute of Government, University of Georgia, in 2008, and is the coeditor (with Byron E. Price) of *Prison Privatization: The Many Facets of a Controversial Industry*, a three-volume set published by Praeger (2012). He has also consulted with the US Environmental Protection (EPA) Agency and the National Association of Schools of Public Affairs and Administration in a program to increase citizen input into the EPA's rulemaking process.

Dr. William Allen Gibson is Batten Professor of Political Science and chair of the Department of Political Science at Virginia Wesleyan College, Virginia Beach, Virginia, where he teaches environmental policy and public administration. He researches and writes about environmental conflict management, environmental policy, and natural resource management. During the 1970s and 1980s, he helped develop the field of alternative environmental conflict management, and helped develop and implement a mediation process for intergovernmental dispute resolution in the 1980s. He has worked during the past fifteen years with grassroots organizations to develop models for local control of natural resources and is currently researching problems faced by local communities in controlling local resources within the current framework of global economic rules, with case studies from several countries in the western hemisphere and Southeast Asia.

Dr. William Marshall Leavitt is an associate professor and MPA director at Old Dominion University. In addition to a previous career with the Denver (CO) Water Authority, Dr. Leavitt has published several articles focused on environmental infrastructure, and has conducted research in the field of social marketing, particularly as it concerns environmental issues.

Shana Campbell Jones, JD, is an environmental lawyer and directs the Virginia Coastal Policy Clinic at William & Mary Law School. Prior to joining William & Mary, Ms. Jones was the executive director of the Center for Progressive Reform (CPR), a nonprofit think tank of fifty legal scholars dedicated to providing commentary and analysis in the area of environmental protection. Ms. Jones has coauthored a series of white papers on improving accountability in the Chesapeake Bay watershed.